Emerging Laser Technologies for High-Power and Ultrafast Science

IOP Series in Coherent Sources, Quantum Fundamentals, and Applications

About the Editor

F J Duarte is a laser physicist based in Western New York, USA. His career has expanded three continents while contributing in the academic, industrial and defense sectors. Duarte is editor/author of 15 laser optics books and sole author of three books: *Tunable Laser Optics, Quantum Optics for Engineers*, and *Fundamentals of Quantum Entanglement*. Duarte has made original contributions in the fields of coherent imaging, directed energy, high-power tunable lasers, laser metrology, liquid and solid-state organic gain media, narrow-linewidth tunable laser oscillators, organic semiconductor coherent emission, *N*-slit quantum interferometry, polarization rotation, quantum entanglement, and space-to-space secure interferometric communications. He is also the author of the generalized multiple-prism grating dispersion theory and pioneered the use of Dirac's quantum notation in *N*-slit interferometry and classical optics. His contributions have found applications in numerous fields, including astronomical instrumentation, dispersive optics, femtosecond laser microscopy, geodesics, gravitational lensing, heat transfer, laser isotope separation, laser medicine, laser pulse compression, laser spectroscopy, mathematical transforms, nonlinear optics, polarization optics, and tunable diode laser design. Duarte was elected Fellow of the Australian Institute of Physics in 1987 and Fellow of the Optical Society of America in 1993. He has received various recognitions, including the *Paul F Foreman Engineering Excellence Award* and the *David Richardson Medal* from the Optical Society.

Coherent Sources, Quantum Fundamentals, and Applications

Since its discovery, the laser, has found innumerable applications from astronomy to… zoology. Subsequently, we have also become familiar with additional sources of coherent radiation such as the free electron laser, optical parametric oscillators, and coherent interferometric emitters. The aim of this book Series in Coherent Sources, Quantum Fundamentals, and Applications is to explore and explain the physics and technology of widely applied sources of coherent radiation and to match them with utilitarian and cutting-edge scientific applications. Coherent sources of interest are those that offer advantages in particular emission characteristics areas such as broad tunability, high spectral coherence, high energy, or high power. An additional area of inclusion are those coherent sources capable of high performance in the miniaturized realm. Understanding of quantum fundamentals can lead to new and better coherent sources and unimagined scientific and technological applications. Application areas of interest include the industrial, commercial, and medical sectors. Also, particular attention is given to scientific applications with a bright future such as coherent spectroscopy, astronomy, biophotonics, space communications, space interferometry, quantum entanglement, and quantum interference.

Publishing benefits
Authors are encouraged to take advantage of the features made possible by electronic publication to enhance the reader experience through the use of colour, animation and video, and incorporating supplementary files in their work.

Do you have an idea of a book that you'd like to explore?
For further information and details of submitting book proposals, see iopscience.org/books or contact Ashley Gasque at ashley.gasque@iop.org.

Emerging Laser Technologies for High-Power and Ultrafast Science

Edited by
François Légaré
Institut national de la recherche scientifique Centre Énergie Matériaux Télécommunications, Varennes, Québec, Canada

IOP Publishing, Bristol, UK

ISBN 978-0-7503-2536-3 (ebook)
ISBN 978-0-7503-2534-9 (print)
ISBN 978-0-7503-2537-0 (myPrint)
ISBN 978-0-7503-2535-6 (mobi)

DOI 10.1088/978-0-7503-2536-3

Version: 20210701

IOP ebooks

British Library Cataloguing-in-Publication Data: A catalogue record for this book is available from the British Library.

Published by IOP Publishing, wholly owned by The Institute of Physics, London

IOP Publishing, Temple Circus, Temple Way, Bristol, BS1 6HG, UK

US Office: IOP Publishing, Inc., 190 North Independence Mall West, Suite 601, Philadelphia, PA 19106, USA

Contents

Preface x

Editor biography xiii

List of contributors xiv

1 High energy, high average power ytterbium lasers 1-1
Jorge J Rocca, Yong Wang, Han Chi, Cory Baumgarten, Brendan A Reagan, Kristian Dehne and Carmen S Menoni

1.1 Introduction 1-1
1.2 Yb-doped gain media 1-2
1.2.1 Yb:YAG at cryogenic temperatures 1-5
1.3 Thin disk lasers 1-6
1.3.1 High energy thin disk regenerative amplifiers 1-8
1.3.2 Thin disk multi-pass amplifiers 1-9
1.4 Cryogenically-cooled Yb:YAG amplifiers 1-11
1.4.1 100 mJ pre-amplifier 1-12
1.4.2 High repetition rate, 1.5 J amplifier 1-12
1.4.3 1.1 J, 1 kHz repetition laser picosecond Yb:YAG laser 1-14
1.4.4 Frequency-doubled 1 J Yb:YAG laser with 1 kW average power 1-17
1.5 Status and prospects 1-21
References 1-23

2 High-performance, thulium-doped, ultrafast fiber lasers 2-1
Christian Gaida, Martin Gebhardt, Cesar Jauregui and Jens Limpert

2.1 Introduction 2-1
2.2 Fundamentals, challenges and hidden scaling potential of ultrafast thulium-doped fiber lasers 2-3
2.2.1 Thulium-doped fused silica as a gain material 2-4
2.2.2 Detrimental propagation effects in ambient air 2-7
2.2.3 The hidden performance scaling potential of 2 μm ultrafast fiber lasers 2-10
2.3 High peak power femtosecond thulium-doped fiber laser 2-13
2.4 High average power, ultrafast thulium-doped fiber lasers 2-17
2.5 Future prospects of frequency conversion driven by high-performance, ultrafast fiber laser systems at 2 μm wavelength 2-21
2.5.1 High harmonic generation (HHG) into the water window 2-22

2.5.2 High-power source in the fingerprint mid-IR spectral region 2-23
2.5.3 High flux THz radiation 2-24
2.6 Conclusion 2-24
References 2-25

3 Transition metal doped zinc selenide infrared lasers for ultrafast and intense field science 3-1

Yi Wu, Esben Witting Larsen, Fangjie Zhou, Lifeng Wang, Alphonse Marra, Daniel Schade, Jialin Li and Zenghu Chang

3.1 Introduction 3-1
3.2 Material properties 3-4
3.3 Thermal lensing 3-6
3.4 Seed pulse generation 3-7
3.4.1 Intra-pulse difference frequency generations 3-7
3.4.2 Optical parametric amplifiers 3-10
3.5 Dispersion control 3-10
3.6 CPA experimental demonstration 3-13
3.7 Gain calculations 3-16
3.8 Post-compression and filamentation for the generation of single-cycle mid-infrared pulses 3-17
3.9 Future perspectives 3-19
Acknowledgments 3-20
References 3-21

4 100 kHz tunable mid-IR ultrafast sources for high intensity applications 4-1

Yoann Pertot, Nicolas Thiré, Raman Maksimenka, Olivier Albert and Nicolas Forget

4.1 Introduction 4-1
4.2 Architecture 4-3
4.3 Generation of a broadband mid-IR seed from a narrowband picosecond pulse 4-5
4.3.1 Optical Kerr effect: self-focusing and self-phase modulation 4-5
4.3.2 Plasma generation: multi-photon ionization 4-7
4.3.3 Linear dispersion and other effects 4-8
4.3.4 Experimental results 4-8
4.3.5 Generation of visible seed 4-9

4.4 Design of a tunable amplification chain 4-10
4.4.1 Concept of parametric amplification 4-11
4.4.2 Pre-amplifier for a tunable source and generation of a CEP-stable seed 4-12
4.4.3 Periodically poled crystals for high amplification stages 4-13
4.4.4 Low gain amplification stage 4-14
4.4.5 Compatibility with a CEP-stable output for strong field applications 4-16
4.5 Dispersion management and compression 4-17
4.5.1 The use of an acousto-optic programmable dispersive filter 4-17
4.5.2 Compression setup for a multi-mode OPCPA system 4-18
4.6 Conclusion 4-19
References 4-20

5 Advances of ultraviolet light sources: towards femtosecond pulses in the few-cycle regime **5-1**
Vincent Wanie, Lorenzo Colaizzi, Andrea Cartella, Andrea Trabattoni and Francesca Calegari

5.1 Introduction 5-1
5.2 Generation of femtosecond UV pulses 5-3
5.2.1 UV generation in crystals 5-6
5.2.2 UV generation in gas cells 5-8
5.2.3 UV generation in fibers and capillaries 5-11
5.2.4 Characterization of few-fs UV pulses 5-13
5.3 Applications in ultrafast molecular science 5-15
5.3.1 Time resolution 5-15
5.3.2 Bandwidth 5-18
5.3.3 Polarization state and future prospects 5-20
5.4 Conclusion 5-22
References 5-23

Preface

High power, ultrafast laser technologies are now used for a wide variety of applications both in fundamental and applied research, touching many scientific disciplines including physics, chemistry, materials science, and biology, but also for societal applications including health, environmental, security, and defense. Ultrafast lasers have enabled major technological advances and are now widely used in industry.

Since 1985, the year when the chirped pulse amplification (CPA) technique was introduced by Strickland and Mourou [1], the peak power of ultrafast lasers has continuously grown reaching recently 10 PW [2], and intensities approaching 10^{23} W cm^{-2} are now available [3]. For more details, here is a recent review [4]. This technology is mostly based on the Titanium:Sapphire (Ti:Sa) amplification gain medium providing pulses of few tens of femtoseconds (fs) at 800 nm center wavelength. For millijoule (mJ) class laser systems, the pulse duration can be further reduced down the single-cycle regime (2.66 fs at 800 nm) using the hollow core fiber approach that was introduced 25 years ago by Nisoli and co-workers [5]. For sub-PW systems, recent progress with thin plate compression provides a route to reach sub-10 fs pulse duration and boost the peak power through a very efficient approach [6].

Over the last two decades, Ti:Sa technology has been the workhorse for ultrafast scientists enabling major breakthroughs that cannot all be discussed in a preface. One amazing outcome of this technology has been the generation of secondary sources from the THz to the hard x-ray spectral range, as well as high energy particles (electrons, ions, and neutrons) using compact laser-driven accelerator schemes. Among many outcomes, in 2021 we are celebrating the twentieth anniversary of the generation of isolated attosecond pulses, which has given rise to the field of attosecond science [7, 8]. Since these light pulses and particles are fully synchronized, sophisticated pump-probe experiments enable the tracking of complex ultrafast dynamics in matter such as attosecond electronic motion in atoms, molecules, and solids [9, 10].

In this context, scientists using ultrafast laser systems want to have access to more average power for generating secondary sources with higher brightness, including high repetition rates, over various regions of the electromagnetic spectrum to drive nonlinear processes that are wavelength dependent such as THz generation and high harmonic generation, to tunable sources to perform spectroscopy, as well as to shorter pulses to improve the temporal resolution of pump-probe experiments. With high repetition rate high average power lasers, one can compensate the low conversion efficiency of nonlinear processes used to generate secondary sources at extreme wavelengths. This book has been written to provide examples of recent achievements in ultrafast optics dealing with these requirements at the forefront of laser technologies.

Chapter 1 discusses the most recent achievements in terms of high energy, high average power laser systems, reporting the use of Ytterbium:YAG gain medium to generate picosecond pulses with kW of average power at kHz repetition rate for

Joule energy per pulse. These performances combine the highest pulse energy and average power to date, and will enable scaling by orders of magnitude the brightness of secondary sources driven by high energy pulses.

Motivated by nonlinear processes that are wavelength dependent, there are major efforts to develop ultrafast laser systems operating at longer wavelengths. In this context, chapters 2–4 discuss different strategies to achieve this goal. Chapter 2 presents the most recent development for high energy, high repetition rate fiber laser systems delivering ultrashort pulses at 2 μm center wavelength. Using thulium-doped fused-silica as a gain material, and benefitting from large core fibers, average power and peak power available nowadays have already reached the same as that obtained with the well-established Ytterbium fiber laser technology. This remarkable achievement paves a roadmap to provide 2 μm laser systems with 100 GW peak power (10 mJ with 100 fs pulse duration) at kW of average power (100 kHz).

Chapter 3 presents recent breakthroughs regarding the development of CPA based on chromium doped zinc selenide gain material (Cr^{2+}:ZnSe) delivering multi-mJ ultrashort pulses at 2.5 μm at kHz repetition rate. To seed the amplifier, the authors present a unique scheme to generate carrier-envelope phase (CEP) stable broadband pulses, resulting in the shortest pulse duration obtained from a Cr^{2+}:ZnSe laser amplifier (44 fs). With filamentation in air for spectral broadening, the bandwidth to generate 17.5 fs pulses is demonstrated. Such a laser system will be ideal to generate isolated attosecond pulses over the entire water window spectral range, even reaching the L-edges of ferromagnetic materials.

In chapter 4, a recent laser technology enabling high repetition rate (100 kHz) tunable narrowband and broadband laser pulses in the infrared and mid-infrared spectral range is presented. Using optical parametric chirped pulse amplification (OPCPA) pumped by high average power Yb:YAG lasers, tens of Watts of average power are available at long wavelengths. This technology, commercialized by Fastlite, capitalizes on their proprietary device called the Dazzler, an acousto-optic programmable dispersive filter that enables to manage the dispersion in the OPCPA to maximize efficiency and reach Fourier transform limited pulse duration over the tunable range. Furthermore, this unique device enables one to switch instantaneously between two modes of operation, i.e., tunable narrow spectral bandwidth for spectroscopic applications and ultrashort CEP stable mid-IR pulses.

Chapter 5 describes the efforts being made to reach the few-optical-cycle regime for ultraviolet pulses. The main route to achieve this task is to exploit low dispersion gas media for frequency up-conversion of Ti:Sa and Ytterbium:YAG laser sources. UV generation methods to obtain few-fs (sub 3-fs) pulse duration are discussed, providing state-of-the-art UV sources as a key ingredient in the field of ultrafast UV spectroscopy. Applications to study in real time the ultrafast response of photo-excited states of matter are discussed by the authors.

To the readers, I really hope that you will enjoy reading this book as much as I did. Even though I have been working in this research field for 20 years, I am always amazed by the progress that is made and by laser scientists, through their ingenuity and hard work, who push the limits and specification of ultrafast laser technologies. As a community, we can envision performances that will enable breakthroughs in fundamental and applied research, thus enabling innovations and new laser based technologies for societal applications.

References

[1] Strickland D and Mourou G 1985 Compression of amplified chirped optical pulses *Opt. Commun.* **55** 447–9
[2] Doria D *et al* 2020 Overview of ELI-NP status and laser commissioning experiments with 1 PW and 10 PW class-lasers *J. Instrum.* **15** C0903
[3] Yoon J W *et al* 2019 Achieving the laser intensity of 5.5 × 1022 W cm^{-2} with a wavefront-corrected multi-PW laser *Opt. Express* **17** 20413
[4] Danson C N *et al* 2019 Petawatt and exawatt lasers worldwide *High Power Laser Sci. Eng.* **7** e54
[5] Nisoli M *et al* 1996 Generation of high energy 10 fs pulses by a new pulse compression technique *Appl. Phys. Lett.* **68** 2793
[6] Mironov S Y *et al* 2020 Thin plate compression of a sub-petawatt Ti:Sa laser pulses *Appl. Phys. Lett.* **116** 241101
[7] Drescher M *et al* 2001 X-ray pulses approaching the attosecond frontier *Science* **291** 1923–7
[8] Hentschel M *et al* 2001 Attosecond metrology *Nature* **414** 509–13
[9] Krausz F and Ivanov M 2009 Attosecond physics *Rev. Mod. Phys.* **81** 163–234
[10] Young L *et al* 2018 Roadmap of ultrafast X-ray atomic and molecular physics *J. Phys. B: At. Mol. Opt. Phys* **51** 032003

Editor biography

François Légaré

Institut national de la recherche scientifique, Centre Énergie Matériaux Télécommunications, Varennes, Québec, Canada

François Légaré, PhD in chemistry from the Université de Sherbrooke (2004), has joined the Institut national de la recherche scientifique—Énergie Matériaux Télécommunications centre in 2006. Since 2013, he has been the director of the Advanced Laser Light Source user facility.

The general goal of his research is to use nonlinear optical phenomena to develop novel ultrafast light sources and approaches to image and control matter.

Légaré is a Fellow and senior member of the Optical Society (2017) and a member of The College of New Scholars, Artists and Scientists of the Royal Society of Canada (2017). He was awarded the Herzberg medal from the Canadian Association of Physics (2015) and the Rutherford Memorial Medal in physics of the Royal Society of Canada (2016). He has contributed to more than 160 articles in peer reviewed journals and has been invited to more than 70 conferences and workshops. He has also contributed to intellectual properties with 4 patents, 4 in national phases, and 1 US provisional. Furthermore, he is fostering the transfer of technologies to Canadian spin-off companies, including *few-cycle* Inc. (founded in 2013, 8 employees) and FemtoGate (founded in 2019). Since 2006, he has enthusiastically supervised and supported 37 internships, 9 MSc and 16 PhD students, and 22 postdoctoral fellows. Currently, his team is composed of 2 MSc, 8 PhD, 7 postdoctoral fellows, 2 research associates, 3 technicians, and 3 research engineers.

List of contributors

Olivier Albert
Fastlite, Pôle Euro 95—Bât D, 165 rue des Cistes, 06600 Antibes, France

Cory Baumgarten
XUV Lasers Inc., Fort Collins, CO 80525, USA

Francesca Calegari
Center for Free-Electron Laser Science CFEL, Deutsches Elektronen-Synchrotron DESY, Notkestraße 85, 22607 Hamburg, Germany

Physics Department, Universität Hamburg, Luruper Chaussee 149, 22761 Hamburg, Germany
The Hamburg Centre for Ultrafast Imaging, Universität Hamburg, Luruper Chaussee 149, 22761 Hamburg, Germany

Andrea Cartella
The Hamburg Centre for Ultrafast Imaging, Universität Hamburg, Luruper Chaussee 149, 22761 Hamburg, Germany

Zenghu Chang
University of Central Florida, Orlando, FL, USA

Han Chi
XUV Lasers Inc., Fort Collins, CO 80525, USA

Lorenzo Colaizzi
Center for Free-Electron Laser Science CFEL, Deutsches Elektronen-Synchrotron DESY, Notkestraße 85, 22607 Hamburg, Germany

Physics Department, Universität Hamburg, Luruper Chaussee 149, 22761 Hamburg, Germany

Kristian Dehne
XUV Lasers Inc., Fort Collins, CO 80525, USA

Nicolas Forget
Fastlite, Pôle Euro 95—Bât D, 165 rue des Cistes, 06600 Antibes, France

Christian Gaida
Active Fiber Systems GmbH, Ernst-Ruska Ring 17, 07745 Jena, Germany

Martin Gebhardt
Institute of Applied Physics, Friedrich-Schiller University Jena, Albert Einstein Str. 15, Beutenberg Campus, 07745 Jena, Germany

Cesar Jauregui
Institute of Applied Physics, Friedrich-Schiller University Jena, Albert Einstein Str. 15, Beutenberg Campus, 07745 Jena, Germany

Esben Larsen
Imperial College London, London SW7 2BW, UK

Jialin Li
University of Central Florida, Orlando, FL, USA

Jens Limpert
Institute of Applied Physics, Friedrich-Schiller University Jena, Albert Einstein Str. 15, Beutenberg Campus, 07745 Jena, Germany

Fraunhofer Institute for Applied Optics and Precision Engineering, Albert Einstein Str. 7, 07745 Jena, Germany

Raman Maksimenka
Fastlite, Pôle Euro 95—Bât D, 165 rue des Cistes, 06600 Antibes, France

Alphonse Marra
University of Central Florida, Orlando, FL, USA

Carmen S Menoni
XUV Lasers Inc., Fort Collins, CO 80525, USA

Brendan A Reagan
XUV Lasers Inc., Fort Collins, CO 80525, USA

Jorge J Rocca
XUV Lasers Inc., Fort Collins, CO 80525, USA

Daniel Schade
Imperial College London, London SW7 2BW, UK

Nicolas Thiré
Fastlite, Pôle Euro 95—Bât D, 165 rue des Cistes, 06600 Antibes, France

Andrea Trabattoni
Center for Free-Electron Laser Science CFEL, Deutsches Elektronen-Synchrotron DESY, Notkestraße 85, 22607 Hamburg, Germany

Lifeng Wang
University of Central Florida, Orlando, FL, USA

Yong Wang
XUV Lasers Inc., Fort Collins, CO 80525, USA

Vincent Wanie
Center for Free-Electron Laser Science CFEL, Deutsches Elektronen-Synchrotron DESY, Notkestraße 85, 22607 Hamburg, Germany

Yi Wu
University of Central Florida, Orlando, FL, USA

Fangjie Zhou
University of Central Florida, Orlando, FL, USA

Chapter 1

High energy, high average power ytterbium lasers

Jorge J Rocca, Yong Wang, Han Chi, Cory Baumgarten, Brendan A Reagan, Kristian Dehne and Carmen S Menoni

1.1 Introduction

Since the invention of the laser, higher pulse energy, greater average power, and shorter pulse duration are constantly evolving frontiers that create new applications and even open entire new fields. The peak power of lasers has increased over the years at a rate that compares with the dramatic increase of the number of transistors in semiconductor chips (Moore's law), at present reaching 10 PW [1]. Impressive increases have also been achieved in pulse energy, with lasers now delivering nanosecond pulses of more than one megajoule [2]. Dramatic reductions in pulse durations have also been realized with laser pulses with hundreds of joules being produced in the femtosecond regime [1]. However, the average power of joule-level ultrashort pulse lasers has, until very recently, remained low. There are important applications in science and technology in which progress depends on the development of compact lasers capable of producing ultrashort laser pulses of high energy at high repetition rate. These applications include coherent and incoherent plasma-based sources of extreme ultraviolet, soft and hard x-rays photons, and the production of gamma rays by inverse Compton scattering. Also, high power, ultrashort pulse lasers operating at repetition rates at and above 1 kHz are required for the implementation of compact particle accelerators based on laser wakefield acceleration for fundamental research and applications. These and other applications motivate the development of high average power, short pulse lasers at institutions worldwide.

Figure 1.1 summarizes the current state-of-the-art of high energy, high average power, short pulse (<10 ps) lasers operating at wavelengths near 1 μm, which is, at present, the wavelength range that has produced the highest average powers.

doi:10.1088/978-0-7503-2536-3ch1

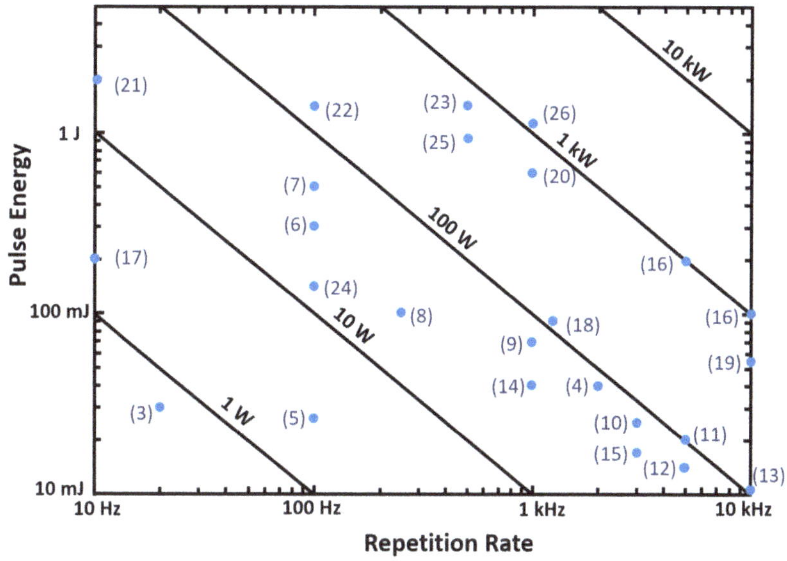

Figure 1.1. Summary of state-of-the-art diode-pumped Yb-doped lasers at $\lambda \approx 1$ μm with high pulse energy (>10 mJ) and repetition rate of up to 10 kHz with pulses compressible to sub-10 ps duration. Each number identifies work listed in the reference section [3–26].

Different approaches, all of them enabled by pumping with laser diodes, have been investigated. The gain geometry has been in the form of slabs, thin disks, or thicker disks operating at cryogenic temperatures. Another approach uses the coherent combination of the output of fiber lasers. Most of these lasers are based on diode-pumping of Yb:YAG, a mature gain material that is readily available and that combines high heat conductivity, large stimulated emission cross section, and relatively broad bandwidth. Yb:YAG picosecond lasers have already reached a pulse energy of 1 J at a repetition rate of 1 kHz, and further increases in both the average power and the pulse energy are expected.

In this chapter we discuss the material properties of Yb:YAG in comparison with other Yb host materials, followed by a summary of its characteristics at cryogenic temperature relative to room temperature. This is followed by a discussion of the state-of-the-art high average power, high energy, thin disk lasers and cryogenic Yb: YAG lasers.

1.2 Yb-doped gain media

The advent of highly efficient, high power semiconductor laser diodes with narrow band emission which better matches the absorption of solid state laser materials than flash lamps has enabled important advances in high power lasers. It has resulted in greatly increased efficiency, larger average power, increased compactness, and reliability. However, a limitation of laser diodes as compared to traditional flash lamps is their relatively low peak power. This requires using gain media with long upper level lifetime that can efficiently store the pump energy. Numerous laser materials with adequate laser upper level lifetime for pumping with laser diodes have been investigated or are under development. The two rare earth ion dopants, neodymium and ytterbium, have so far dominated the development of high power lasers. They are well suited for diode pumping due to the inherent wavelength matching between their absorption bands and that of commercially available high power diodes. Nd^{+3} can be efficiently pumped by λ = 790–900 nm InGaAsP/GaAs or AlGaAs/GaAs laser diodes, and Yb^{+3} by λ = 900–980 nm InGaAs/GaAs laser diodes. Yet Yb^{3+} materials such as Yb:YAG have the advantage of having an upper level lifetime of ~1 ms, which is over four times longer than similar Nd^{3+} materials like Nd:YAG, which has a lifetime of 230 μs. This allows for better pump energy usage, and therefore reduced cost. Furthermore, when pumped at λ = 942 nm Yb: YAG has an absorption bandwidth of 18 nm full width at half maximum (FWHM) [27], which is nearly an order of magnitude greater than the 2 nm absorption bandwidth of Nd:YAG at λ = 808 nm. This relaxes diode linewidth and center wavelength stability requirements.

Figure 1.2 shows the energy level diagrams for Nd^{3+} and Yb^{3+}. In both materials lasing takes place at wavelengths around 1 μm in transitions between the $^4F_{5/2}$ and $^4I_{11/2}$ manifolds in Nd^{3+} and between the $^2F_{5/2}$ and $^2F_{7/2}$ manifolds in Yb^{3+}. A first observation is the different value of the quantum defect between the two materials, i.e., the difference between pump and emission wavelengths. When expressed as the fraction of the lasing wavelength to the pump wavelength, Nd:YAG has a quantum defect of 24% when pumped at 808 nm, while Yb:YAG has a value of 9% when pumped at 942 nm. This is significant because the quantum defect is often the primary direct source of heating in the material. Therefore, minimizing the lasing to pump wavelength difference is key for high power laser operation, where thermal

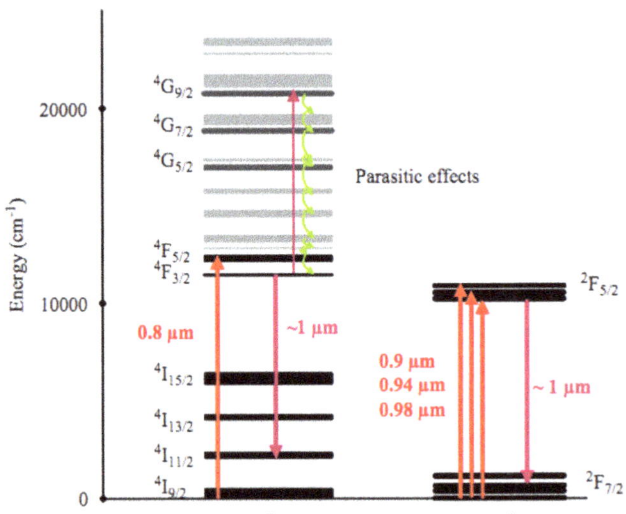

Figure 1.2. The relevant energy levels of Nd^{3+} and Yb^{3+} gain media with typical pump and lasing transitions, as well as parasitic effects. Reprinted from Chenais *et al* [27], Copyright 2006, with permission from Elsevier.

management is one of the major challenges. Yb^{3+} also has a simplified electronic energy level diagram that helps to avoid most parasitic effects such as energy transfer upconversion, excited state absorption, and cross-relaxation, which all occur in Nd-doped media because of the existence of higher energy levels [27]. The main de-excitation paths of these high energy levels are primarily non-radiative and contribute to further thermal loading of the system. These deleterious effects can also depopulate the laser upper level, therefore reducing the population inversion and the gain. Finally, an advantage for the development of ultrafast lasers is the broader laser line emission bandwidth of Yb-doped materials, that is about an order of magnitude broader than of Nd-doped materials, allowing for the generation of shorter transform limited laser pulses. This broader gain bandwidth combined with longer storage lifetime comes with the consequence of lower peak stimulated emission cross section and the corresponding higher saturation fluence. Additionally, the small quantum defect results in a non-negligible thermal population on the lower laser level which increases threshold pumping requirements. However, as we discuss below, these complications are ameliorated at cryogenic operating temperatures. Overall, the Yb^{3+} ion has emerged over Nd^{+3} as a prime candidate for high repetition rate, high power, ultrafast laser development.

Numerous Yb-doped gain media with different host materials exist that exhibit many of the spectroscopic properties discussed above. These materials have long upper level lifetimes, τ_L, and small quantum defects, but differences in material properties make some more suitable than others for high power operation. These materials include Yb:YAG, Yb:KYW, Yb:KGW, Yb:YLF, and Yb:CaF_2. Some of their spectroscopic and mechanical properties are listed in table 1.1.

Several of these materials have been explored as options for high power lasers [33–36]. Hosts such as YLF and CaF_2 are attractive due to their broad emission bandwidth, $\Delta\lambda$, and good thermal conductivities, κ, and could constitute alternatives for high energy femtosecond lasers. These materials have found use in short pulse high power oscillators [37] and low energy amplifiers [3, 13, 38]. Yb:CaF_2 has also been used in diode-pumped high energy amplifiers [39, 40] to generate laser pulses of up to 54 J at repetition rates of less than 1 Hz, compressible to ~100 fs. Yet, its low stimulated emission cross section, σ_e, means it has a low gain, and a high saturation fluence. The saturation fluence of Yb:CaF_2 can exceed 100 J cm^{-2}, which is well

Table 1.1. Spectroscopic and mechanical properties of various Yb-doped materials. From S Chenais *et al* [27]. Other values have also been reported [28–32].

Crystal host	λ_c (nm)	λ_{abs} (nm)	$\Delta\lambda$ (nm)	σ_e (10^{-20} cm^2)	τ_L (ms)	κ (W m^{-1} K^{-1}) (undoped)
YAG	1030	942	9	2.1	0.95	11
KYW	1025	981	24	3	0.3	3.3
KGW	1023	981	20	2.8	0.3	3.3
YLF	1030	959	14	0.81	2.21	4.3
CaF_2	1049	980	70	0.25	2.4	9.7

above the damage threshold of optical coatings and even of the host material itself. This is a significant limitation for its use in the implementation of high average power, high energy amplifiers, where it is necessary to operate above the saturation fluence for efficient energy extraction. In the case of KYW and KGW, their low thermal conductivities have limited their use to mode-locked oscillators [41, 42] and low energy amplifiers [33, 34]. Of the available host materials, Yb:YAG, which has relatively high thermal conductivity and comparatively low saturation fluence, has emerged as the most utilized material for high energy, high repetition rate picosecond laser operation. Yttrium aluminum garnet ($Y_3Al_5O_{12}$), or YAG, has been a popular host material for decades due to its attractive spectroscopic, mechanical, and thermal properties. YAG has a body-center cubic lattice that makes it optically isotropic. However, thermal stress can still cause depolarization. YAG also has a higher fracture strength than most laser materials and is well suited to accept substitution of trivalent rare earth ions. Furthermore, since YAG has been studied as a laser host since the mid-1960s [43] and has found wide use with Nd^{3+} and Yb^{3+} dopants, the fabrication of high quality single crystal and ceramic YAG is mature, making large aperture YAG-based gain media commercially available. In recent decades, Yb:YAG has found wide use in high power picosecond chirped pulse amplification (CPA) lasers using a number of different designs including water-cooled thin disks [14–16], slabs [19, 44], and thicker disks or slabs operating at cryogenic temperatures [23–26]. Cryogenic Yb:YAG is particularly well suited for high pulse energy operation at high repetition rates.

1.2.1 Yb:YAG at cryogenic temperatures

The thermo-optical and mechanical properties of Yb:YAG as a gain medium for high power lasers improve greatly at cryogenic temperature [28, 45–47]. From table 1.1, we see that at room temperature YAG has a thermal conductivity of 11 W m^{-1} K^{-1}, which is better than many of the other host materials listed. However, it is still significantly worse than other host materials like sapphire used in Ti:Sapphire lasers, which has a thermal conductivity of 33 W m^{-1} K^{-1} at room temperature. Nevertheless, when cooled to cryogenic temperatures, the thermal conductivity of undoped YAG increases by nearly a factor of 9 with respect to room temperature. While this advantage is reduced when moderate levels of doping are introduced, a significant improvement of 3–5 times still remains [28]. In addition, as shown in table 1.2, the thermal expansion coefficient as well as the thermo-optic coefficient improve at cryogenic temperatures. The former is reduced by a factor of 4 as the temperature decreases from 300 K to 77 K, while the thermo-optics coefficient decreases by a factor of 7. In addition, the emission spectrum of Yb:YAG is significantly narrower at 77 K, which leads to an increase of the stimulated emission cross section and the corresponding decrease in the saturation fluence to below 2 J cm^{-2} [46]. This decrease in saturation fluence allows for efficient energy extraction without optical damage to the coatings and bulk materials. The associated increase in gain also allows for efficient energy extraction in a low number of passes through the gain media, which helps to avoid high nonlinear phase accumulation. On the

Table 1.2. Optical and mechanical properties of Yb:YAG at room temperature and at liquid nitrogen temperature. The column to the right quantifies the change. Values according to references [28, 45–47].

Yb:YAG properties at room and cryogenic temperature	300 K	77 K	Factor
Thermal conductivity (W m^{-1} K^{-1})	10	90	× 9
Thermo-optic coefficient (10^{-6}/K)	7.8	0.9	× 1/7
Expansion coefficient (10^{-6}/K)	6.14	1.95	× 1/4
Saturation fluence (J cm^{-2})	9.2	1.7	× 1/7

other hand, the narrower spectral bandwidth limits the shortest pulsewidth that can in practice be obtained to a few picoseconds. Therefore, several high energy cryo-cooled Yb:YAG laser system designs that suffer bandwidth narrowing in the amplifier chains have used hybrid approaches that combine either different cryogenically-cooled Yb-doped amplifier materials [18] or cryogenically-cooled power amplifiers with a room temperature first amplification stage where most of the gain takes place [23, 48].

A disadvantage of Yb:YAG's simple two manifold electronic structure and low quantum defect is that the lower laser level is thermally populated at room temperature. From Maxwell–Boltzmann statistics, the fraction of ions in the lower laser level has a population of 5.3% at 300 K due to thermal excitation from the ground state. This makes room temperature Yb:YAG a quasi-three level laser (figure 1.3(left)). Since there is an overlap between the absorption and emission bands, this leads to significant re-absorption and to a reduction of the effective emission bandwidth [28]. Therefore, there is a minimum amount of pump energy required for the system to reach transparency, and a corresponding reduction in the laser efficiency. This changes greatly when Yb:YAG is cooled to cryogenic temperatures, i.e., at ~77 K, liquid nitrogen temperature. At this temperature the fraction of ions in the lower laser level is reduced to 10^{-5}, making Yb:YAG a four level laser system (figure 1.3(right)), in which a population inversion and gain start to build up at low pump energy.

1.3 Thin disk lasers

The thin disk laser concept for the implementation of high average power diode-pumped solid state lasers, which dates back to 1994 [49], is illustrated in figure 1.4. The approach consists in the use of an active medium in the shape of a thin disk, typically 100–300 μm thick, that is cooled through one of the flat faces which is bonded to a water-cooled heat sink. The cooled face is coated with a dielectric multilayer stack that is highly reflective for both the pump and the laser wavelengths, while the opposite side of the disk is coated with an anti-reflection coating at both wavelengths. This geometry, which acts as a mirror with gain, is known as an 'active mirror', which can be used in reflection in an oscillator or amplifier. As the pump spot size is typically much larger than the disk thickness, one-dimensional heat flow results in which the temperature gradients are mainly coaxial to the disk

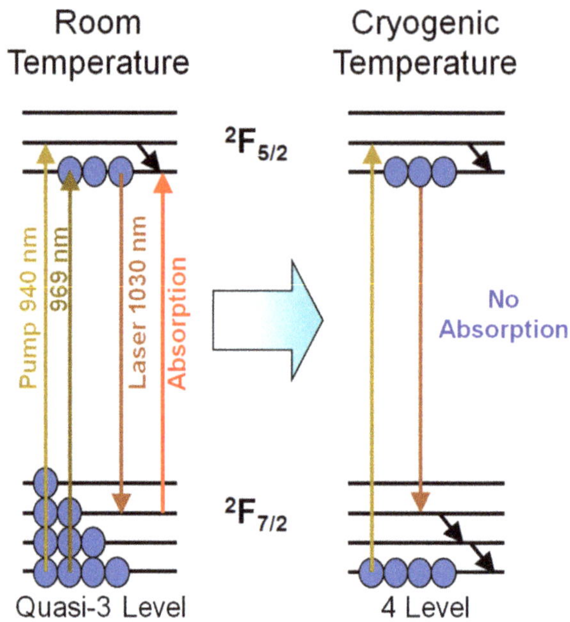

Figure 1.3. Population of the laser lower level in Yb:YAG at room and cryogenic temperature. The laser material changes from a quasi-three level system at room temperature (left) into a four level system at cryogenic temperature (right).

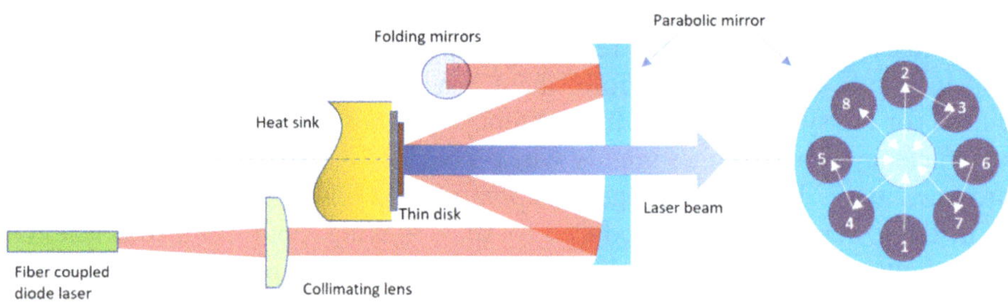

Figure 1.4. (a) Schematic illustration of the thin disk configuration for 16 pump passes. © 2007 IEEE. Reprinted, with permission, from Giesen *et al* [50].

and the laser beam. Therefore, the thermal lens effect can be greatly reduced with respect to an edge-cooled rod geometry, allowing for power scaling. Also, the large surface-to-volume ratio allows for efficient heat dissipation from the active medium into the heat sink, enabling operation at very high power densities.

Because of the small crystal thickness only a small fraction of the pump power is absorbed, requiring multiple passes of the seed beam through the gain medium. Figure 1.4 shows a typical arrangement to increase the number of pump beam passes through the crystal employing a parabolic mirror and a set of prism retro-reflectors. After each pass, the unabsorbed portion of the pump beam is collimated again by the other side of the parabolic mirror that focuses it onto the disk. This re-imaging is repeated until all the positions of the parabolic mirror are covered, the point at which the pump beam can be redirected back to the source, thereby doubling the number of passes through the thin disk. This provides a platform for efficient pumping in which more than 90% of the pump power is absorbed in the disk. The small thickness also results in a small single pass gain that for efficient amplification requires multiples passes either in a resonator of a multi-pass amplifier configuration (figure 1.5(a)), or in a regenerative amplifier (figure 1.5(b)).

1.3.1 High energy thin disk regenerative amplifiers

The thin disk configuration has been used with Yb:YAG to create high average power regenerative amplifiers with output pulse energy of up to 100–200 mJ at repetition rates of 5–10 kHz [16]. This 1 kW average power CPA regenerative amplifier system used the architecture illustrated in figure 1.6(a) [16]. Pulses of 1.3 μJ energy from a Kerr-mode-locked thin disk oscillator are stretched to 1.5 ns duration in a grating stretcher. The stretched pulses are amplified to 1–2 mJ in a first thin disk regenerative amplifier. A following main amplifier is built around two diode-pumped water-cooled Yb:YAG thin disk amplifier heads that constitute the gain media in the ring-type resonator illustrated in figure 1.6(b). CW pump diodes

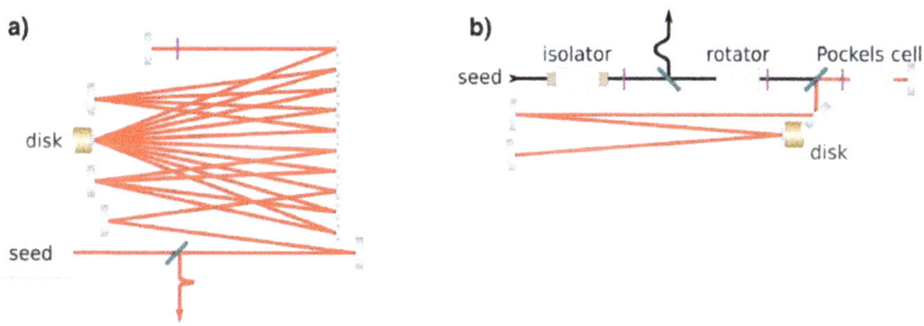

Figure 1.5. Schematic representation of (a) multi-pass amplifier, and (b) thin disk regenerative amplifier. From Saraceno *et al* [51] (2019). With permission of Springer.

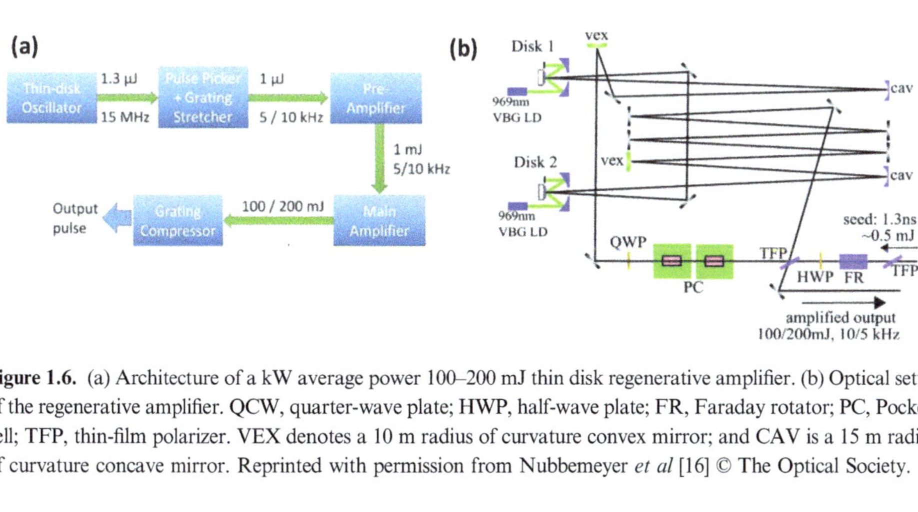

Figure 1.6. (a) Architecture of a kW average power 100–200 mJ thin disk regenerative amplifier. (b) Optical setup of the regenerative amplifier. QCW, quarter-wave plate; HWP, half-wave plate; FR, Faraday rotator; PC, Pockels cell; TFP, thin-film polarizer. VEX denotes a 10 m radius of curvature convex mirror; and CAV is a 15 m radius of curvature concave mirror. Reprinted with permission from Nubbemeyer *et al* [16] © The Optical Society.

operating at wavelengths of 969 nm are used to pump 6.8 mm spots into the 200 μm thick thin disks. The pump diodes are stabilized by volume Bragg gratings to enable direct pumping of the upper laser level via the narrow bandwidth zero phonon line. The resonator mode is symmetric with respect to the position of the two laser disks, ensuring similar mode size in both crystals. The first regenerative amplifier is used to increase the seed pulse energy to avoid chaotic behavior and achieve stable operation of the main amplifier.

As shown in figures 1.7(a) and (b), this system produces average powers exceeding 1 kW with the 100–200 mJ output pulse energies mentioned above. The spectral bandwidth of the pulses seeding the pre-amplifier is 3.5 nm and is reduced to 1.5 nm by gain narrowing in the two subsequent amplifier stages (figure 1.7(b)). The pulses were compressed to 1.08 ps using an efficient dielectric grating compressor. Another thin disk regenerative amplifier implementation has produced uncompressed pulses of slightly higher energy, 300 mJ, at 100 Hz repetition rate, that were subsequently compressed to 1.8 ps FWHM pulses of >240 mJ energy [52].

1.3.2 Thin disk multi-pass amplifiers

Thin Yb:YAG disks have also been used to implement high average power, multi-pass amplifiers which provide a platform to further increase the pulse energy. Outputs with compressed pulses of up to 720 mJ at 1 kHz repetition rate have been reported [20]. The laser system that produced these pulses is schematically illustrated in figure 1.8. It combines a commercially available regenerative amplifier that is based on a ring configuration, figure 1.8(b), with a multi-pass amplifier, both pumped by $\lambda = 940$ nm laser diodes as shown in figure 1.8(c).

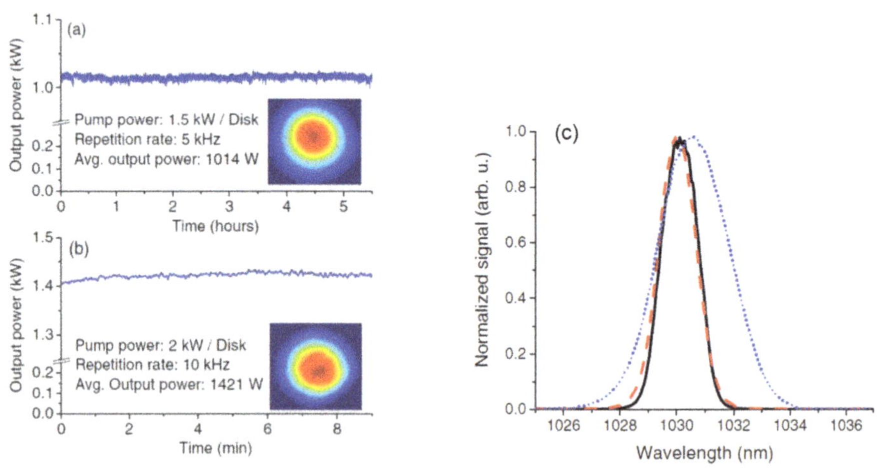

Figure 1.7. (a) Output power of thin disk regenerative amplifiers at 5 kHz repletion rate. (b) Output power at 10 kHz repetition rate achieved over several minutes. The insets show the measured output beam profiles for each repetition rate. (c) Spectra at the output of each of the stages of the regenerative amplifier depicted in figure 1.6: stretched seed pulse (blue dotted line), 1 mJ pre-amplifier output (red dashed line), and output of the main amplifier (black solid line). Reprinted with permission from Nubbemeyer *et al* [16] © The Optical Society.

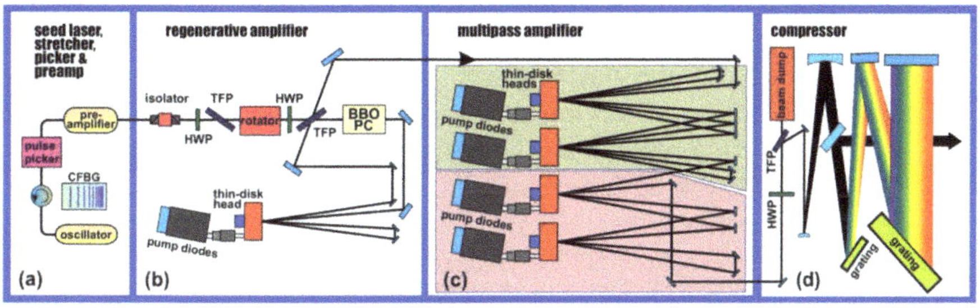

Figure 1.8. Thin disk laser system producing 720 mJ compressed pulses at 1 kHz repetition rate based on a thin disk regenerative amplifier followed by a multi-pass amplifier that uses four thin disks' amplifier heads. TFP, thin-film polarizer; HWP, half-wave plate; BBO PC, BBO-based Pockels cell. Reprinted with permission from Herkommer *et al* [20] © The Optical Society.

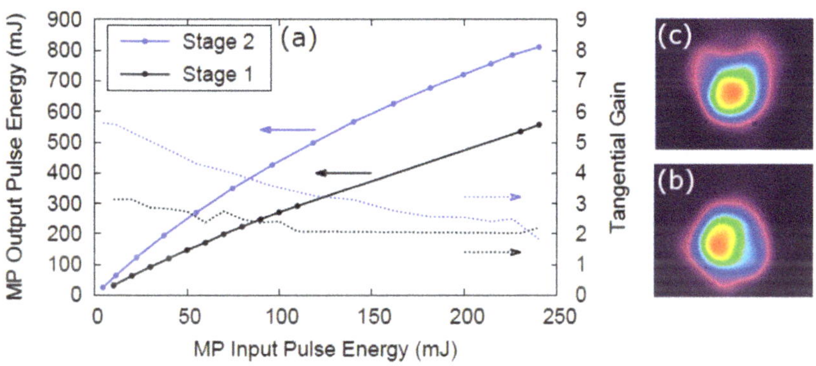

Figure 1.9. (a) Output pulse energy amplification characteristics of the two stages of the thin disk multi-pass amplifier of figure 1.8 as a function of seed pulse energy. The pump power is kept constant. (b) and (c) show the beam profiles after stage 1 and 2, respectively. Reprinted with permission from Herkommer *et al* [20] © The Optical Society.

The multi-pass amplifier is arranged in two stages, each made of two amplifier heads. A total average accumulated pump power of >9 kW is absorbed in the four water-cooled amplifier heads. The thermal lens at the operation point is compensated to obtain near-collimated propagation. The approach allows guiding of the seed beam over the thin disks multiple times with adjustable beam diameters using mirrors which are selected depending on the evolving beam size and divergence. Figure 1.9 shows the measured pulse output energy of each amplification stage as a function of seed pulse energy, maintaining constant pump power. In the first stage of this multi-pass amplifier, 240 mJ pulses from the regenerative pre-amplifier are amplified in two laser heads to 550 mJ in seven passes. In the second stage, the pulse energy is further increased to 800 mJ by four additional disk reflections. The pulses exiting the multi-pass amplifier are compressed into 720 mJ pulses of 0.92 ps duration in a dielectric grating compressor. The M^2 at the output of the amplifier was measured to be 1.89/2.32 in the major/minor axis.

1.4 Cryogenically-cooled Yb:YAG amplifiers

A different approach for the development of high energy, high average power amplifiers takes advantage of the improved thermo-optical and laser properties of Yb:YAG at cryogenic temperatures. The design architecture and performance of a compact $\lambda = 1.03$ μm CPA laser developed at Colorado State University based on cryogenically-cooled active mirror Yb:YAG amplifiers that produces pulses of up to 1.5 J at a repetition rate of 500 Hz is described below. The output pulses are compressed in vacuum to <5 ps in a dielectric grating compressor, yielding pulse energies of up to 1 J. A more recent implementation has produced 1.1 J picosecond pulses at 1 kHz repetition rate [26].

The approach relies on cryogenically-cooled Yb:YAG active mirrors. Although in these amplifiers the thickness of the gain medium (a few millimeters) is significantly larger than in the thin disk lasers, they share some of the same advantages. The geometry is still sufficiently thin to develop mostly longitudinal thermal gradients, reducing thermal lensing as compared to radially-cooled gain geometries. It also limits the maximum temperature difference within the gain region while its moderate diameter-to-length ratio reduces amplified spontaneous emission (ASE) depletion of the amplifier storage energy [53]. The improved thermal parameters at cryogenic temperatures compensate for the larger thickness of this geometry. Compared with thin disk lasers, the increased absorption reduces the number of passes necessary to efficiently absorb the pump, simplifying the design.

A block diagram showing the layout of the 500 Hz repetition rate, 1 J, picosecond laser system is shown in figure 1.10(a). The laser front-end is composed of a semiconductor saturable absorber mirror (SESAM) mode-locked Yb:KYW oscillator, a grating pulse stretcher, and a room temperature Yb:YAG regenerative amplifier. Since

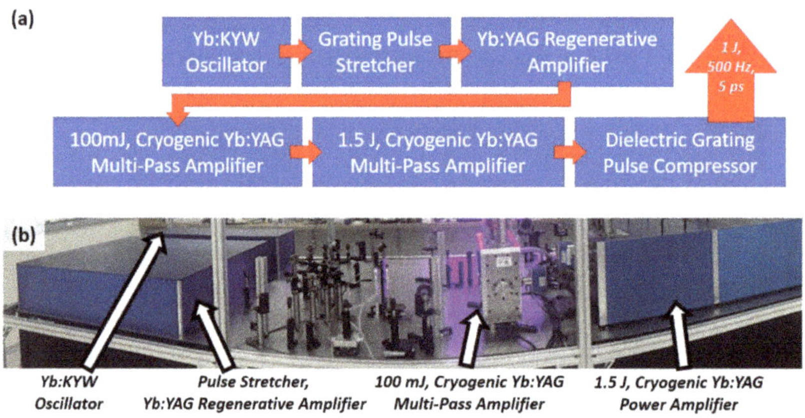

Figure 1.10. (a) Diagram of the laser system based on a high power, cryogenically-cooled Yb:YAG amplifier. The output of a diode-pumped, mode-locked bulk Yb:KYW oscillator is stretched by a grating stretcher and amplified to the millijoule level by a room temperature Yb:YAG regenerative amplifier. These seed pulses are subsequently amplified by a chain of two high power cryogenic amplifiers, and are compressed in vacuum by a dielectric grating pair. (b) Panoramic photograph of the high energy, high average power laser. Reproduced with permission from Reagan *et al* [54] © Cambridge University Press.

Yb:KYW is well suited to pumping by $\lambda = 980$ nm laser diodes, the crystal is pumped by a 30 W laser diode operating at that wavelength coupled to a 200 μm core fiber. The femtosecond pulses produced at 55 MHz repetition rate by the Yb:KYW oscillator are stretched by a 1740 line/mm grating to ~270 ps duration. The first stage of amplification is a room temperature Yb:YAG regenerative amplifier. This amplifier has the highest gain of any in the system, approximately 10^6, and is therefore where pulses suffer the most gain narrowing. Therefore, the choice of room temperature operation allows for a broader bandwidth [46]. Because of the mismatch between the center wavelength and the bandwidth of room temperature and cryogenic Yb:YAG, the stretcher is adjusted to select a spectral bandwidth that matches the region of peak gain in the subsequent cryogenic amplifiers while striking for a compromise between maximum pulse energy and broad spectral bandwidth at the exit of the final amplifier. Pulses leaving the grating stretcher seed the regenerative amplifier cavity via a thin-film polarizer (TFP). The active medium is a 0.5 mm thick, 10%-at. doped Yb:YAG active mirror soldered to a water-cooled copper heat sink. The Yb:YAG disk is pumped by ~200 W pulses of 200 μs duration from a fiber-coupled $\lambda = 940$ nm semiconductor laser diode. The stretched pulses suffer gain narrowing during amplification, and their pulse duration becomes 270 ps. These pulses can be compressed to ~2 ps durations, but their bandwidth is reduced when the wavelength is tuned to match the peak gain of the subsequent cryogenic amplifiers. The millijoule-level pulses are first amplified to about 100 mJ in a cryo-cooled pre-amplifier, and finally to the Joule level in the main multi-pass amplifier.

1.4.1 100 mJ pre-amplifier

The layout of the 100 mJ-level amplifier is shown in figure 1.11(a). A 5 mm thick, 2%-at. Yb:YAG active mirror is mounted on a cryogenic cooling manifold enclosed in a vacuum chamber. The highly reflective face of the active mirror is cooled by flowing cryogenic liquid at 77 K [23]. The Yb:YAG active mirror is pumped by $\lambda = 940$ nm, 500 μs duration pulses produced by two 400 W peak power fiber-coupled laser diode modules. A pump spot diameter of ~3.5 mm was used, and the pump radiation is absorbed in a single pass (one reflection).

The seed pulses from the laser front-end make four passes through the amplifier. Figure 1.11(b) shows the measured pulse energy exiting the amplifier as a function of pump energy at 500 Hz repetition rate. As can be seen from this plot, an amplified pulse energy of 100 mJ (50 W average power) is obtained when pumped with 360 mJ at 500 Hz repetition rate. These pulses have a bandwidth of 0.43 nm FWHM (figure 1.11(c)) and were compressed to 3.8 ps FWHM duration as shown by the SHG autocorrelation of figure 1.11(d).

1.4.2 High repetition rate, 1.5 J amplifier

The Joule-level amplifier is shown in figure 1.12(a). Two Yb:YAG active mirrors are mounted on a single cryogenic cooling head. The active mirrors are composites consisting of a 30 mm × 30 mm × 2 mm thick 3%-at. Yb:YAG crystal optically bonded on the four lateral faces to an absorbing Cr^{4+}:YAG cladding to reduce feedback of transversely emitted ASE and avoid parasitic oscillations. The front face

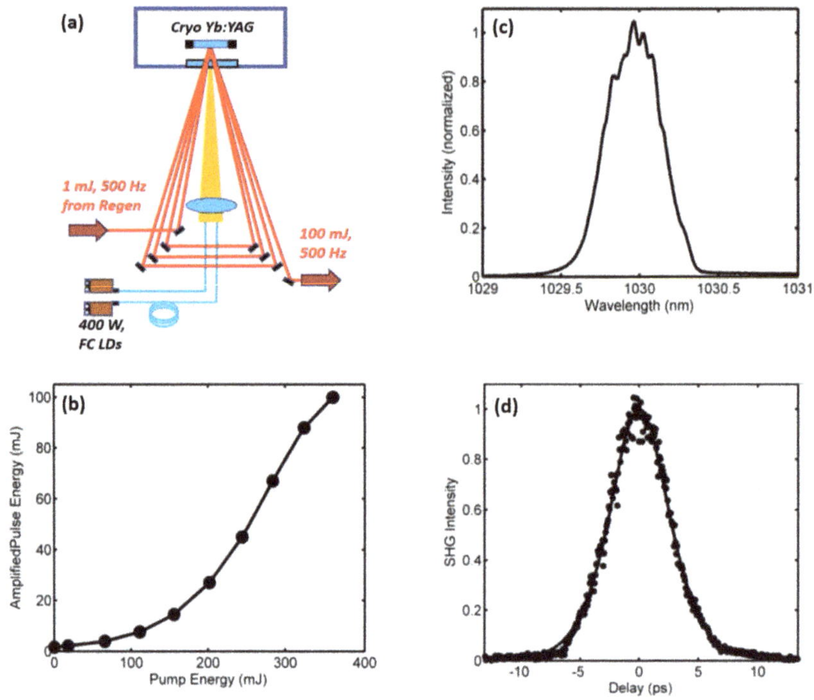

Figure 1.11. (a) Schematic layout of a 100 mJ, 500 Hz cryogenic amplifier. (b) Measured pulse energy at 500 Hz repetition rate as a function of total pump pulse energy. (c) Spectrum of the amplified pulses showing a bandwidth of 0.43 nm FWHM. (d) Autocorrelation of the compressed 100 mJ pulses at 500 Hz repetition rate. The $sech^2$ fit shown corresponds to a pulse duration of 3.9 ps FWHM. Reproduced with permission from Reagan *et al* [54] © Cambridge University Press.

of the assembly is bonded to an undoped YAG cap to further reduce ASE and to add mechanical stability. As shown in figure 1.12(a), each active mirror is pumped by pulses of 500 μs duration from a $\lambda = 940$ nm 6 kW laser diode stack. The pump beam diameter is ≈16 mm, and is double-passed to achieve >95% absorption. Seed pulses are injected into the amplifier through a TFP and are directed to pass through each active mirror twice. After double-passing a quarter-wave plate, the seed pulses pass back through the amplifier along the same path, achieving a total of four passes through each active mirror before being ejected from the amplifier by reflecting off the TFP. The amplifier occupies a table area of just over 0.5 m^2.

Figure 1.13 shows the performance characteristics of this amplifier operating at 500 Hz repetition rate. The laser output pulse energy as a function of total pump energy is displayed in figure 1.13(a). An output energy of 1.5 J (750 W average power) is obtained with an optical-to-optical efficiency of 37%. Figure 1.13(b) shows the measured output pulse energy stability over 30 min of continuous operation at 500 Hz repetition rate. A mean pulse energy of 1.4 J was obtained with an RMS deviation of 0.75%. Figure 1.13(c) shows a measured M^2 parameter of ~1.3 on both axes. The good beam quality in the far field is illustrated in figure 1.13(d). The pulses were compressed with about 70% efficiency by a pair of 1740 mm^{-1} dielectric

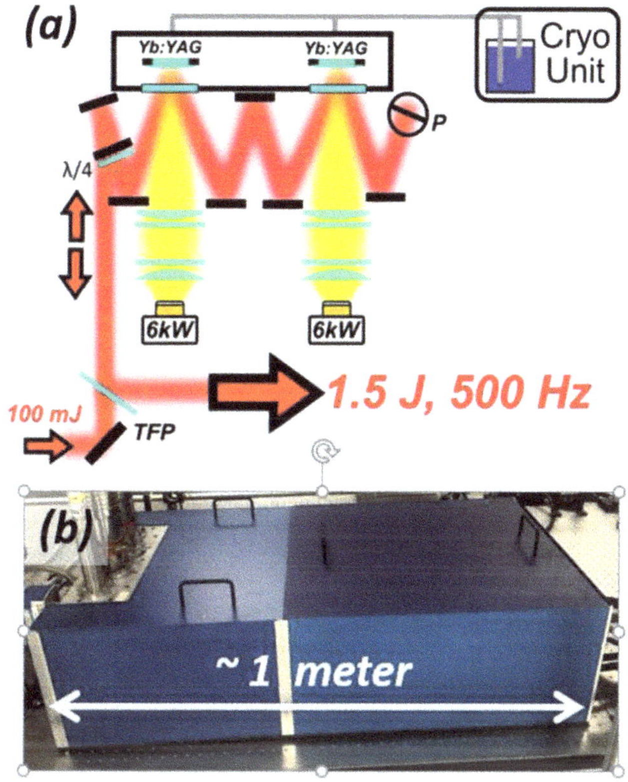

Figure 1.12. (a) Layout of the high repetition rate, 1.5 J amplifier. (b) Photo of the enclosed amplifier, which occupies a table space of just over 1 m × 0.5 m. TFP, thin-film polarizer; λ/4, quarter-wave plate; P, periscope; 6 kW laser diode stacks. Reproduced with permission from Reagan *et al* [54] © Cambridge University Press.

diffraction gratings. The resulting pulses had an energy of 1 J and a duration of ~5 ps at 500 Hz repetition rate, generating an average power of 0.5 kW.

1.4.3 1.1 J, 1 kHz repetition laser picosecond Yb:YAG laser

A different version of this system resulted in the first demonstration of a picosecond laser emitting >1 J pulses at 1 kHz repetition rate [26]. This cryogenically-cooled system generates pulses with >1.2 J energy at a repetition rate of 1 kHz, corresponding to an average power of 1.26 kW. After compression with high efficiency dielectric gratings, 1.1 J pulses with a duration of 4.5 ps were generated with good beam quality and high shot-to-shot stability. A schematic of this kW average power CPA picosecond laser system is illustrated in figure 1.14. It consists of a room temperature front-end that generates 2 mJ seed pulses followed by two cryogenic amplification stages and a vacuum dielectric grating compressor.

Differing from the 500 Hz system [23], the room temperature, thin disk regenerative Yb:YAG amplifier is pumped at its zero phonon line by 500 μs pulses from a 969 nm wavelength 60 W fiber-coupled laser diode. An 8-pass pump geometry forms a well-overlapped pump area of 1 mm diameter with a pump

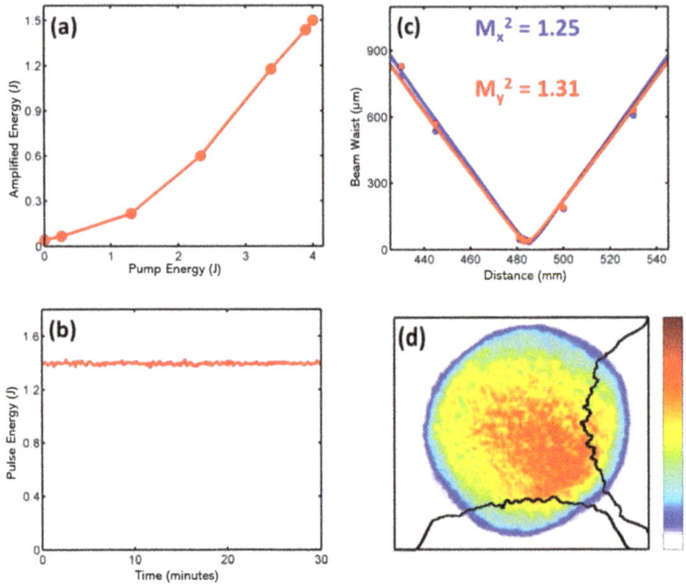

Figure 1.13. (a) Measured pulse energy exiting the main amplifier at 500 Hz repetition rate as a function of combined pump energy. (b) Amplified pulse energy over 30 min of continuous operation at 500 Hz repetition rate. A mean energy of 1.4 J was measured with an RMS fluctuation of 0.75%. (c) M^2 measurement in two orthogonal directions of the amplified output producing 1.4 J at 500 Hz repetition rate. (d) Far-field beam image of the full power output. Reproduced with permission from Reagan *et al* [54] © Cambridge University Press.

absorption of over 95%. A plano-concave cavity allows for four passes through the gain medium per round trip. The pre-amplifier consists of two active mirrors based on 5 mm thick 2%-at. Yb:YAG crystals mounted into a single evacuated chamber where their back surfaces are cooled by flowing liquid nitrogen. The number of active mirrors in this pre-amp was increased to two to better distribute the thermal load and to allow for a longer effective gain medium. A total of seven passes achieves the desired amplification to 100 mJ pulse energy at 1 kHz repetition rate. The first of the two active mirrors is pumped by imaging the output of a 400 W, $\lambda = 940$ nm, fiber-coupled diode laser into a 4 mm diameter spot. The laser pulses are amplified in five passes through the first crystal and are subsequently further amplified in two passes through the second active mirror that is pumped into a similar spot by two 400 W fiber-coupled laser diodes. For both crystals, the pump pulse duration is set at 450 μs. The ~100 mJ pulses are passed through an 8 mm diameter serrated aperture (SA) to achieve a nearly flat-top beam profile and pass through a Faraday rotator before injection into the final dual crystal cryogenic amplifier. Each crystal in this final amplifier stage is pumped at $\lambda = 940$ nm by a 6 kW diode array. The pump diodes generate 380 μs duration pulses at 1 kHz that are shaped to form nearly flat-top 16-mm-diameter pump profiles onto each active mirror. The temperature at the center of the pump region was measured to remain below 130 K (figure 1.15) using a spectroscopic technique [55]. After the seed pulses pass twice through each of the two active mirrors, a periscope is used to spatially rotate the beam 90 degrees and change its height. The beam is sent back to the two active

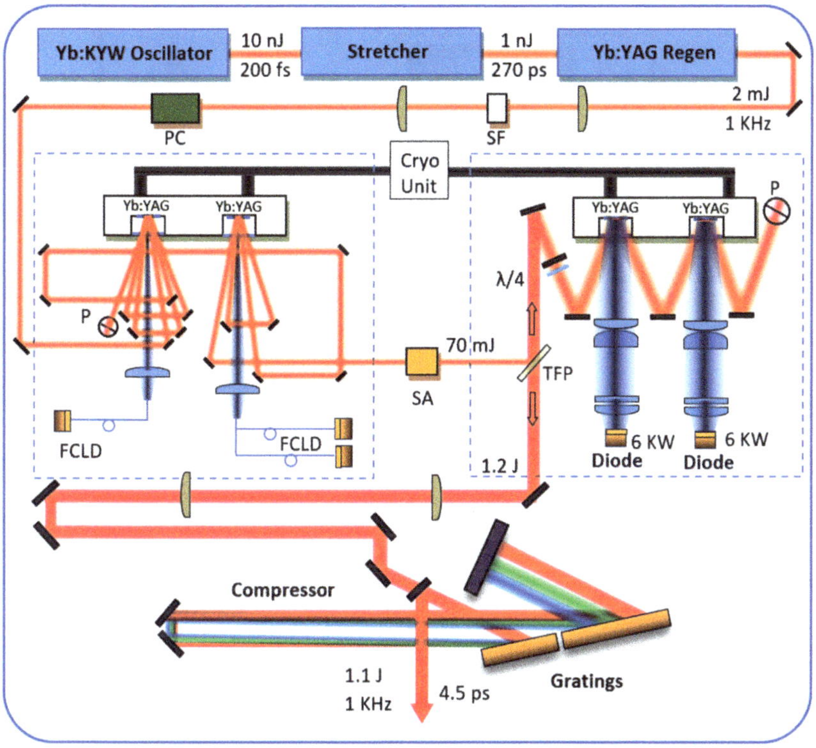

Figure 1.14. Schematic of the diode-pumped high energy CPA laser system. Regen, regenerative amplifier; SF, spatial filter; PC, Pockels cells; FCLD, fiber-coupled laser diodes; SA, serrated aperture; TFP, thin-film polarizer; P, periscope. Reprinted with permission from Wang *et al* [26] © The Optical Society.

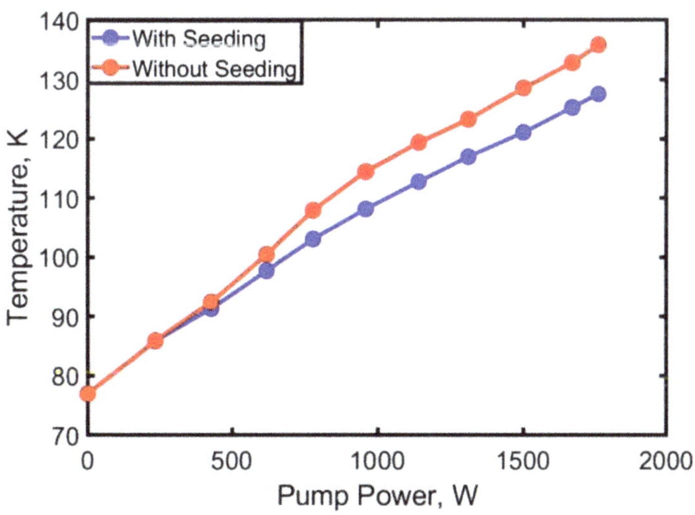

Figure 1.15. Measured temperature of one Yb:YAG active mirror in the final amplifier at the center of the pump with and without seeding vs. pump power. Reprinted with permission from Wang *et al* [26] © The Optical Society.

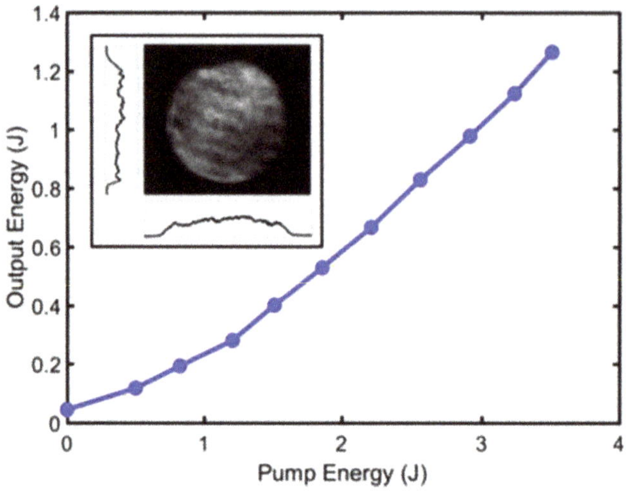

Figure 1.16. Measured laser pulse energy exiting the final amplifier as a function of total pump energy at 1 kHz repetition rate. A maximum energy of 1.26 J was obtained with a total pump energy of 3.5 J incident on the Yb:YAG active mirror. The inset shows the beam profile at 1 kHz, 1.26 kW average power. Reprinted with permission from Wang *et al* [26] © The Optical Society.

mirrors for the third and fourth pass. After the first four passes a quarter-wave plate combined with a 0-degree high reflector (HR) mirror rotates the polarization 90 degrees, sending the beam back through the same path to the TFP to exit the amplifier.

Figure 1.16 shows the measured output pulse energy as a function of the final amplifier total pump energy at 1 kHz repetition rate. A maximum 1.26 J pulse energy (1.26 kW average power) was obtained with a total pump pulse energy of 3.5 J. The optical-to-optical conversion efficiency is ~36%. The pulse-to-pulse energy stability is characterized by a standard deviation of 0.6%. As also shown by the inset in figure 1.16, the 1.26 kW average power output beam has a uniform flat-top profile. The beam is characterized by a measured $M^2 = 1.32$ and 1.39 on the horizontal and vertical directions, respectively (figure 1.17).

Figure 1.18(a) shows the measured spectra before the pre-amplifier and after the main amplifier. A 0.35 nm FWHM bandwidth is obtained at the output of the final amplifier stage, showing a minimal amount of gain narrowing, which supports a transform limited pulse duration of 3.9 ps ($sech^2$). Following amplification the beam is magnified to a diameter of ~40 mm and pulses are compressed with ~90% efficiency as shown by the SHG autocorrelator $sech^2$ fit in figure 1.18(b) corresponding to a pulse duration of 4.48 ps FWHM. The demonstrated 1.2 kW average power amplifier chain also provides the opportunity to generate nanosecond lasers pulses by substituting the short pulse laser front-end with one generating seed pulses of nanosecond duration.

1.4.4 Frequency-doubled 1 J Yb:YAG laser with 1 kW average power

The output of $\lambda = 1.030$ μm Yb:YAG lasers can be efficiently frequency-doubled to drive high pulse energy femtosecond sources based on Ti:Sapphire or optical

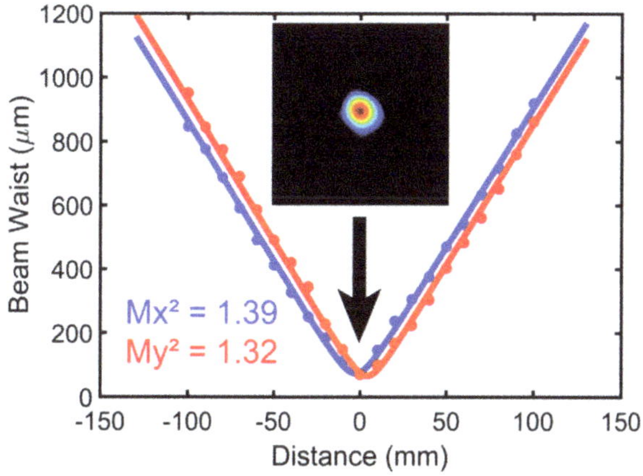

Figure 1.17. M^2 measurement of the 1.2 J, 1 kHz repetition rate output beam. The inset image shows the far-field beam profile at focus. Reprinted with permission from Wang *et al* [26] © The Optical Society.

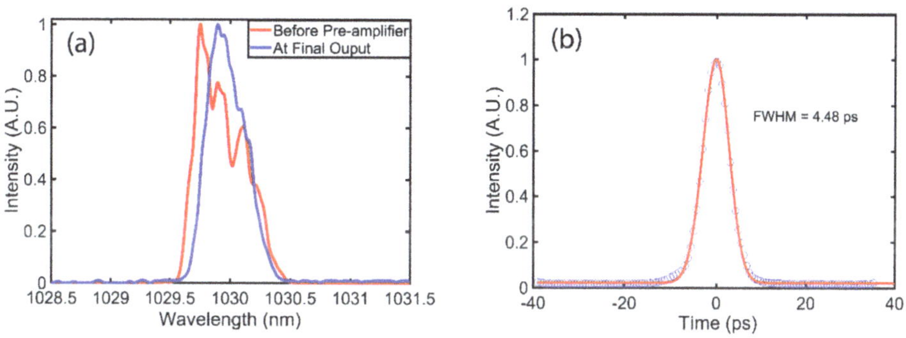

Figure 1.18. (a) Measured spectra at the input of the pre-amplifier (red) and the output of the final amplifier (blue). The pulses have an amplified bandwidth of 0.35 nm FWHM. (b) Second-harmonic autocorrelation after pulse compression. The solid trace is the sech^2 fit with a width corresponding to a pulse duration of 4.48 ps. Reprinted with permission from Wang *et al* [26] © The Optical Society.

parametric chirped pulse amplifiers (OPCPA) at kHz repetition rates. Most of the progress achieved in the development of high average power green lasers based on Yb:YAG has been limited to the generation of μJ to mJ-level pulses at repetition rates ranging from hundreds of kHz to MHz [56, 57]. For example, 370 W average power at $\lambda = 515$ nm was achieved by the generation of 7 μJ pulses at 50 MHz repetition rate [56]. An average power of 1.4 kW at the same wavelength was achieved with 4.8 mJ pulses at 300 kHz repetition rate [57]. In comparison, the repetition rate and average power of lasers generating Joule-level green pulses has remained considerably lower until recently. SHG in the DIPOLE laser based on a cryogenically-cooled Yb:YAG laser has produced 5.6 J pulses at 10 Hz repetition rate, an average power of 56 W [58]. SHG in yttrium calcium oxyborate (YCOB) of pulses from the Yb:S-FAP Mercury laser produced 31.7 J at 10 Hz repetition rate (317 W average power) [59]. A 1 kHz cryogenically-cooled Yb:YAG laser similar to

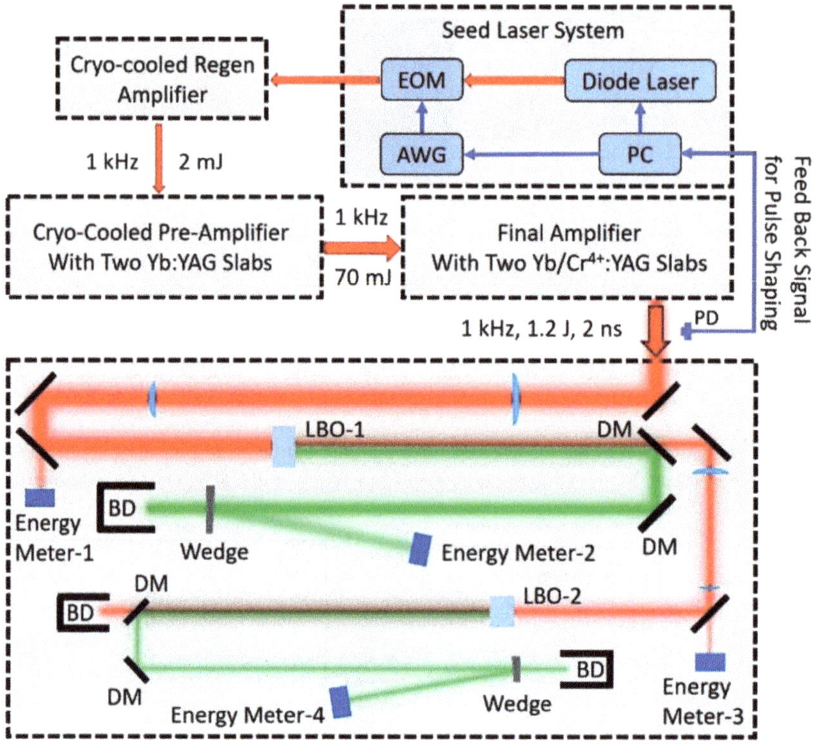

Figure 1.19. Schematic of the high average power, 1 J, green laser setup. The high power λ = 515 nm laser system includes a seed pulse laser front-end that generates pulses of arbitrary shape, a chain of diode-pumped Yb:YAG cryogenic amplifiers, and two sequentially placed LBO SHG units. DM, dichroic mirror; BD, beam dumper; PD, photodiode. Reprinted with permission from Chi *et al* [60] © The Optical Society.

that described above has been frequency-doubled in LBO to generate 1 J green pulses and 1 kW average power [60].

This frequency-doubled Yb:YAG laser is schematically illustrated in figure 1.19. Infrared laser pulses of 1.2 J energy are produced with 1.2 kW average power and are frequency-doubled in two LBO crystals. Temporally square pulses were generated to maintain the intensity impinging into the doubling crystals at the optimum value for frequency doubling. Because the temporal profile of the pulses injected into the amplifier chain is re-shaped by gain saturation as they are amplified to the Joule level, the seed pulses are generated by an arbitrary-waveform laser, which is programmable to generate a square pulse shape at the end of the amplifier chain. A schematic of the setup used to generate seed pulses of arbitrary shape is shown in the upper part of figure 1.19. The seed laser comprises a 2 W peak power $\lambda = 1030$ nm diode laser, an electro-optic modulator (EOM), an arbitrary-waveform generator (AWG), and a processing unit. The seed semiconductor laser is temperature tuned to operate at the wavelength corresponding to high gain in the following chain of Yb:YAG cryogenic amplifiers. A photodiode is placed at the output of the amplifier chain to monitor the final pulse shape and provide the feedback necessary

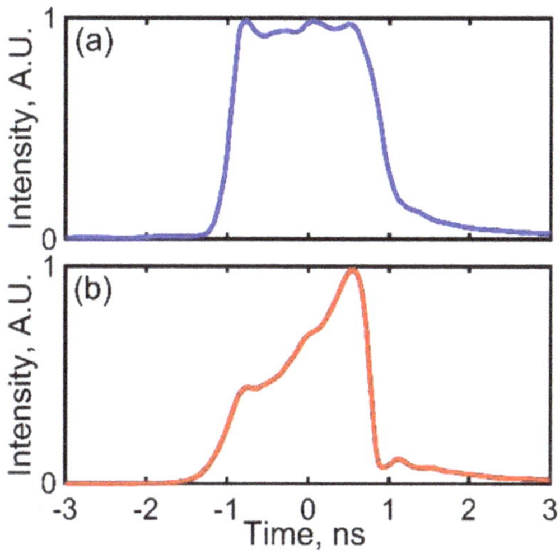

Figure 1.20. Temporal profiles of the (a) output pulse from the final amplifier, and (b) the seed pulse measured before the regenerative amplifier. Reprinted with permission from Chi *et al* [60] © The Optical Society.

to adjust the input signal of the programmable AWG. The adjusted signal is then again applied to the EOM. This feedback process repeats until the desired pulse shape is achieved at the output of the amplifier chain. Figure 1.20(a) shows the measured temporal profile of the amplified 1.2 J square pulse, along with that corresponding to the shaped seed pulse before the cryogenically-cooled regenerative amplifier (figure 1.20(b)). While only the square pulse shape relevant to SHG is presented here, other pulse shapes can be generated with the same setup for other applications.

The shaped seed pulses were amplified to an energy of >1.2 J at 1 kHz repetition rate by a chain of cryogenically-cooled Yb:YAG amplifiers similar to that described above, but in which the regenerative amplifier stage is also cryogenically cooled. LBO crystals cut at $\theta = 90°$ and $\phi = 13.6°$ for Type-I phase matching were placed at the output of the final amplification stage for upconversion into λ = 515 nm laser light. To make the most efficient use of the fundamental beam, two LBO crystals mounted on temperature-controlled copper holders at 300 K were set up in series as shown in figure 1.19. A Keplerian telescope was used to image the output of the Yb:YAG amplifier chain with selected demagnification onto the first LBO crystal (13 mm thick). The second-harmonic beam was separated from the fundamental by a sequence of two dichroic mirrors. The unconverted λ = 1030 nm beam was imaged into a second LBO crystal. Figure 1.21 shows the λ = 515 nm laser pulse energy and SHG conversion efficiency as a function of the λ = 1030 nm pulse energy obtained with the 13-mm-thick LBO crystal for a 14 mm diameter beam with a fluence of up to 0.78 J cm^{-2}. The pulse energy reaches 0.94 J, corresponding to an average power of 940 W with a shot-to-shot standard deviation of 2.9% and a conversion efficiency of 78%. The unconverted beam was upconverted in the second LBO crystal, which generated additional >100 mJ pulses at λ = 515 nm. The resulting total λ = 515 nm average power reaches 1.04 kW (1.04 J pulses at 1 kHz repetition rate).

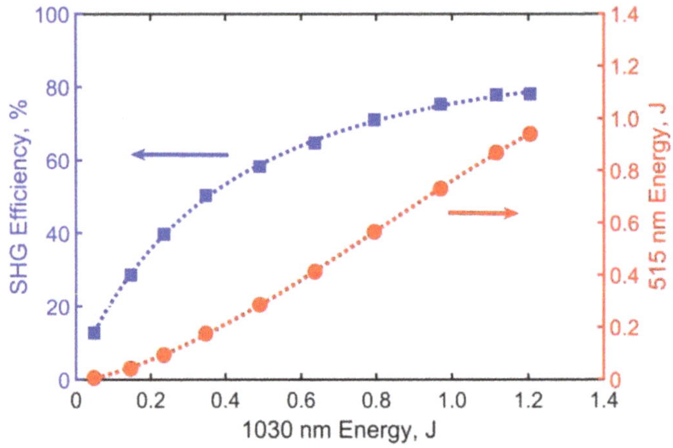

Figure 1.21. SHG efficiency (blue) and SHG output energy (red) vs. fundamental pulse energy obtained with a 13-mm-thick LBO crystal at 1 kHz repetition rate. Reprinted with permission from Chi *et al* [60] © The Optical Society.

Figures 1.22(a) and (b) show that both the $\lambda = 515$ nm, 0.94 J and the secondary 100 mJ beams display flat-top profiles with good uniformity. The second-harmonic beam is characterized by a measured $M^2 = 1.40$ and 1.32 on the horizontal and vertical directions, respectively (figure 1.22(c)).

An even higher conversion efficiency of 89% was obtained with a 10-mm-thick LBO crystal when 0.65 J fundamental wavelength pulses were downsized to achieve a fluence of 0.83 J cm^{-2}, as shown in figure 1.23.

These SHG results will enable the generation of high energy laser pulses of femtosecond duration at 1 kHz repetition rate. In addition, the generation of Joule-level nanosecond pulses of arbitrary shape from a kW average power Yb:YAG laser system will benefit other applications such as heating tailored plasmas for extreme ultraviolet light generation.

1.5 Status and prospects

Yb:YAG picosecond lasers have already reached a pulse energy of 1 J at a repetition rate of 1 kHz. Further increases in both the average power and the pulse energy are expected. Yb:YAG is a mature, thoroughly studied material that is broadly available with high optical quality. It has the advantages of relatively long upper level lifetime that allows for efficient pumping with commercial high power laser diodes. It also has high thermal conductivity and small quantum defect, both of which ease the challenges of thermal management. In addition, Yb:YAG has a relatively broad spectral bandwidth that allows for the generation of picosecond pulses using chirped pulse amplification. These properties that make Yb:YAG attractive for the development of high average power, high energy ultrashort pulse lasers, further improve at cryogenic temperatures. When used in an active mirror geometry either in a thin disk geometry or in thicker gain medium geometries at cryogenic temperature, Yb:YAG enables the generation of Joule-level output pulses at record average powers. Compressed pulses of

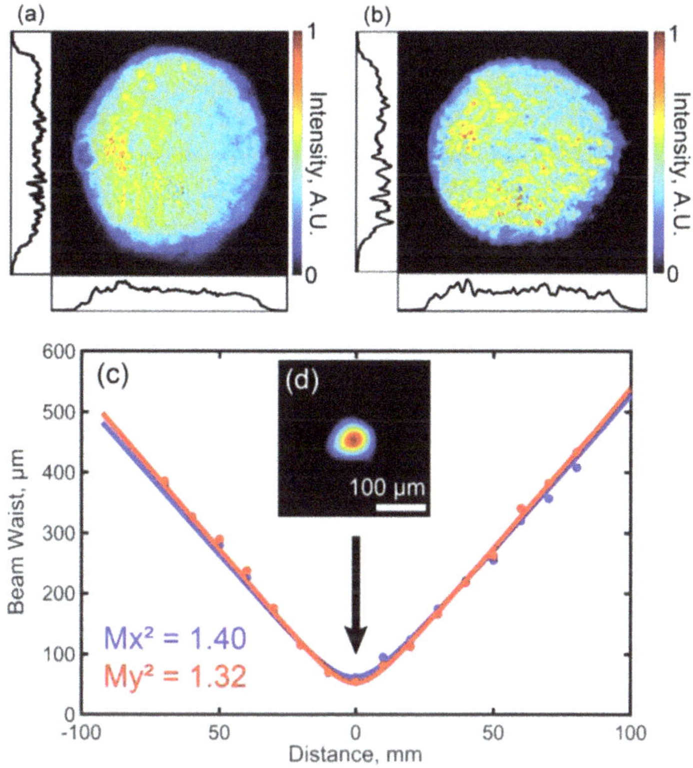

Figure 1.22. Intensity profiles of the $\lambda = 515$ nm beams of pulse energy (a) 0.94 J, and (b) 100 mJ at 1 kHz repetition rate; (c) M^2 measurement of the $\lambda = 515$ nm primary beam, and (d) beam profile at the focal spot. Reprinted with permission from Chi *et al* [60] © The Optical Society.

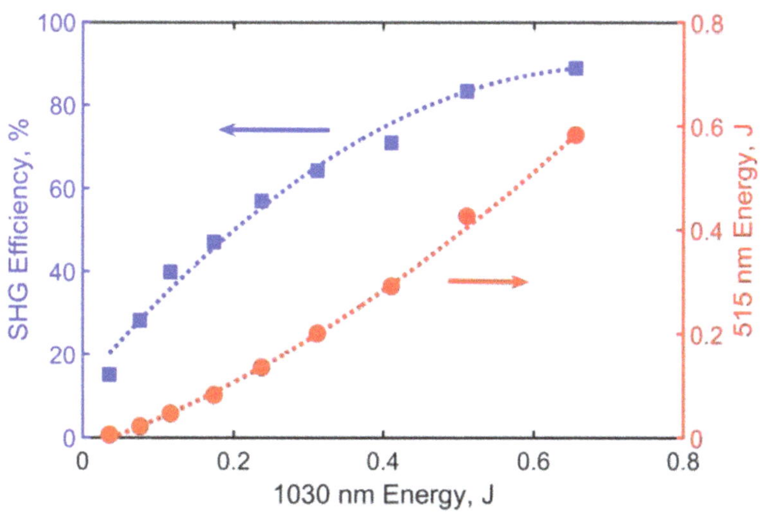

Figure 1.23. SHG efficiency (blue) and SHG output energy (red) vs. fundamental energy obtained with a 10-mm-thick LBO crystal at 1 kHz repetition rate. Reprinted with permission from Chi *et al* [60] © The Optical Society.

720 mJ energy and 0.92 ps duration at 1 kHz repetition rate have been reported for a laser based on thin disks. A laser system based on cryogenically-cooled Yb:YAG has produced 1.1 J pulses of 4.5 ps duration at a 1 kHz repetition rate, an average power of 1.1 kW, which is a record for Joule-level picosecond lasers. These Joule-level high repetition rate lasers have already been used in applications that include pumping plasma-based soft x-ray lasers and producing filaments in the atmosphere to guide lightning. Future trends include the further increase in both the average power and the pulse energy, and the reduction of the pulse duration. Multi-Joule outputs at several kHz can be expected. Finally, while the pulse duration of these lasers has been bandwidth limited to about 1 ps, techniques for spectral broadening of high energy pulses are feasible and offer a path to femtosecond operation [61].

References

[1] Danson C N *et al* 2019 Petawatt and exawatt class lasers worldwide *High Power Laser Sci. Eng.* **7** 54

[2] Moses E I, Boyd R N, Remington B A, Keane C J and Al-Ayat R 2009 The National Ignition Facility: ushering in a new age for high energy density science *Phys. Plasmas* **16** 13

[3] Kawanaka J, Yamakawa K, Nishioka H and Ueda K 2003 30-mJ, diode-pumped, chirped-pulse Yb:YLF regenerative amplifier *Opt. Lett.* **28** 2121–23

[4] Hong K H, Gopinath J T, Rand D, Siddiqui A M, Huang S W, Li E, Eggleton B J, Hybl J D, Fan T Y and Kartner F X 2010 High-energy, kHz-repetition-rate, ps cryogenic Yb:YAG chirped-pulse amplifier *Opt. Lett.* **35** 1752–54

[5] Papadopoulos D N, Pellegrina A, Ramirez L P, Georges P and Druon F 2011 Broadband high-energy diode-pumped Yb:KYW multipass amplifier *Opt. Lett.* **36** 3816–18

[6] Tummler J, Jung R, Stiel H, Nickles P V and Sandner W 2009 High-repetition-rate chirped-pulse-amplification thin-disk laser system with joule-level pulse energy *Opt. Lett.* **34** 1378–80

[7] Jung R, Tummler J, Nubbemeyer T and Will I 2016 Thin-disk ring amplifier for high pulse energy *Opt. Express* **24** 4375–81

[8] Zapata L E, Lin H, Calendron A L, Cankaya H, Hemmer M, Reichert F, Huang W R, Granados E, Hong K H and Kartner F X 2015 Cryogenic Yb:YAG composite-thin-disk for high energy and average power amplifiers *Opt. Lett.* **40** 2610–13

[9] Chang C L *et al* 2015 High-energy, kHz, picosecond hybrid Yb-doped chirped-pulse amplifier *Opt. Express* **23** 10132–44

[10] Metzger T, Schwarz A, Teisset C Y, Sutter D, Killi A, Kienberger R and Krausz F 2009 High-repetition-rate picosecond pump laser based on a Yb:YAG disk amplifier for optical parametric amplification *Opt. Lett.* **34** 2123–25

[11] Fattahi H, Alismail A, Wang H C, Brons J, Pronin O, Buberl T, Vamos L, Arisholm G, Azzeer A M and Krausz F 2016 High-power, 1-ps, all-Yb:YAG thin-disk regenerative amplifier *Opt. Lett.* **41** 1126–29

[12] Rand D A *et al* 2011 Picosecond pulses from a cryogenically cooled, composite amplifier using Yb:YAG and Yb:GSAG *Opt. Lett.* **36** 340–42

[13] Miller D E, Zapata L E, Ripin D J and Fan T Y 2012 Sub-picosecond pulses at 100 W average power from a Yb:YLF chirped-pulse amplification system *Opt. Lett.* **37** 2700–2

[14] Novak J *et al* 2016 Thin disk amplifier-based 40 mJ, 1 kHz, picosecond laser at 515 nm *Opt. Express* **24** 5728–33

[15] Fischer J, Heinrich A C, Maier S, Jungwirth J, Brida D and Leitenstorfer A 2016 615 fs pulses with 17 mJ energy generated by an Yb:thin-disk amplifier at 3 kHz repetition rate *Opt. Lett.* **41** 246–49
[16] Nubbemeyer T *et al* 2017 1 kW, 200 mJ picosecond thin-disk laser system *Opt. Lett.* **42** 1381–84
[17] Klingebiel S, Wandt C, Skrobol C, Ahmad I, Trushin S A, Major Z, Krausz F and Karsch S 2011 High energy picosecond Yb:YAG CPA system at 10 Hz repetition rate for pumping optical parametric amplifiers *Opt. Express* **19** 5357–63
[18] Morrissey F X, Fan T Y, Miller D E and Rand D 2017 Picosecond kilohertz-class cryogenically cooled multistage Yb-doped chirped pulse amplifier *Opt. Lett.* **42** 707–10
[19] Schmidt B E, Hage A, Mans T, Legare F and Woerner H J 2017 Highly stable, 54 mJ Yb-InnoSlab laser platform at 0.5 kW average power *Opt. Express* **25** 17549–55
[20] Herkommer C *et al* 2020 Ultrafast thin-disk multipass amplifier with 720 mJ operating at kilohertz repetition rate for applications in atmospheric research *Opt. Express* **28** 30164–73
[21] Furch F J, Reagan B A, Luther B M, Curtis A H, Meehan S P and Rocca J J 2009 Demonstration of an all-diode-pumped soft x-ray laser *Opt. Lett.* **34** 3352–54
[22] Reagan B A *et al* 2014 1 Joule, 100 Hz repetition rate, picosecond CPA laser for driving high average power soft x-ray lasers *Conf. on Lasers and Electro-Optics (CLEO)* (San Jose, CA: IEEE)
[23] Baumgarten C, Pedicone M, Bravo H, Wang H C, Yin L, Menoni C S, Rocca J J and Reagan B A 2016 1 J, 0.5 kHz repetition rate picosecond laser *Opt. Lett.* **41** 3339–42
[24] Curtis A H, Reagan B A, Wernsing K A, Furch F J, Luther B M and Rocca J J 2011 Demonstration of a compact 100 Hz, 0.1 J, diode-pumped picosecond laser *Opt. Lett.* **36** 2164–66
[25] Zapata L E, Schweisthal S, Thesinga J, Zapata C, Schust M, Yizhou L, Calendron A-L, Pergament M and Kaertner F X 2019 Joule-class 500 Hz cryogenic Yb:YAG chirped pulse amplifier *Conf. on Lasers and Electro-Optics Europe and European Quantum Electronics Conf. 2019* (Munich: Optical Society of America)
[26] Wang Y, Chi H, Baumgarten C, Dehne K, Meadows A R, Davenport A, Murray G, Reagan B A, Menoni C S and Rocca J J 2020 1.1 J Yb:YAG picosecond laser at 1 kHz repetition rate *Opt. Lett.* **45** 6615–18
[27] Chenais S, Druon F, Forget S, Balembois F and Georges P 2006 On thermal effects in solid-state lasers: the case of ytterbium-doped materials *Prog. Quantum Electron.* **30** 89–153
[28] Aggarwal R L, Ripin D J, Ochoa J R and Fan T Y 2005 Measurement of thermo-optic properties of $Y_3Al_5O_{12}$, $Lu_3Al_5O_{12}$, YAIO(3), $LiYF_4$, $LiLuF_4$, BaY_2F_8, $KGd(WO_4)(2)$, and $KY(WO_4)(2)$ laser crystals in the 80–300 K temperature range *J. Appl. Phys.* **98** 14
[29] Calendron A L, Cankaya H and Kartner F X 2014 High-energy kHz Yb:KYW dual-crystal regenerative amplifier *Opt. Express* **22** 24752–62
[30] Siebold M, Bock S, Schramm U, Xu B, Doualan J L, Camy P and Moncorge R 2009 Yb: CaF2 a new old laser crystal *Appl. Phys.* B **97** 327–38
[31] Kuleshov N V, Lagatsky A A, Podlipensky A V, Mikhailov V P and Huber G 1997 Pulsed laser operation of Yb-doped $KY(WO_4)(2)$ and $KGd(WO_4)(2)$ *Opt. Lett.* **22** 1317–19
[32] Koechner W 2006 *Solid State Laser Engineering* (Berlin: Springer)
[33] Ogawa K, Akahane Y, Aoyama M, Tsuji K, Tokita S, Kawanaka J, Nishioka H and Yamakawa K 2007 Multi-millijoule, diode-pumped, cryogenically-cooled Yb:$KY(WO_4)(2)$ chirped-pulse regenerative amplifier *Opt. Express* **15** 8598–602

[34] Liu H H, Nees J and Mourou G 2002 Directly diode-pumped Yb:KY(WO_4)(2) regenerative amplifiers *Opt. Lett.* **27** 722–24
[35] Caracciolo E, Kemnitzer M, Guandalini A, Pirzio F, Agnesi A and Au J A D 2014 High pulse energy multiwatt Yb:$CaAlGdO_4$ and Yb:CaF_2 regenerative amplifiers *Opt. Express* **22** 19912–18
[36] Fan T Y, Ripin D J, Aggarwal R L, Ochoa J R, Chann B, Tilleman M and Spitzberg J 2007 Cryogenic Yb^{3+}-doped solid-state lasers *IEEE J. Sel. Top. Quantum Electron.* **13** 448–59
[37] Machinet G *et al* 2013 High-brightness fiber laser-pumped 68 fs-2.3 W Kerr-lens mode-locked Yb:CaF_2 oscillator *Opt. Lett.* **38** 4008–10
[38] Pugzlys A *et al* 2009 Multi-mJ, 200-fs, cw-pumped, cryogenically cooled, Yb,Na:CaF_2 amplifier *Opt. Lett.* **34** 2075–77
[39] Albach D, Loeser M, Siebold M and Schramm U 2019 Performance demonstration of the PENELOPE main amplifier HEPA I using broadband nanosecond pulses *High Power Laser Sci. Eng.* **7** 9
[40] Hornung M *et al* 2016 54 J pulses with 18 nm bandwidth from a diode-pumped chirped-pulse amplification laser system *Opt. Lett.* **41** 5413–16
[41] Brunner F *et al* 2002 240-fs pulses with 22-W average power from a mode-locked thin-disk Yb:KY(WO_4)(2) laser *Opt. Lett.* **27** 1162–64
[42] Brunner F *et al* 2000 Diode-pumped femtosecond Yb:KGd(WO_4)(2) laser with 1.1-W average power *Opt. Lett.* **25** 1119–21
[43] Geusic J E, Marcos H M and Vanuitert L G 1964 Laser oscillations in Nd-doped yttrium aluminum yttrium gallium + gadolinium garnets (continuous operation of $Y_3A1_5O_{12}$, pulsed operation of $Y_3Ga_5O_{15}$ + $Gd_3Ga_5O_{12}$ rm temperature) *Appl. Phys. Lett.* **4**(10) 182
[44] Russbueldt P, Mans T, Weitenberg J, Hoffmann H D and Poprawe R 2010 Compact diode-pumped 1.1 kW Yb:YAG Innoslab femtosecond amplifier *Opt. Lett.* **35** 4169–71
[45] Wynne R, Daneu J L and Fan T Y 1999 Thermal coefficients of the expansion and refractive index in YAG *Appl. Opt.* **38** 3282–84
[46] Dong J, Bass M, Mao Y L, Deng P Z and Gan F X 2003 Dependence of the Yb^{3+} emission cross section and lifetime on temperature and concentration in yttrium aluminum garnet *J. Opt. Soc. Am.* B **20** 1975–79
[47] Slack G A and Oliver D W 1971 Thermal conductivity of garnets and phonon scattering by rare-earth ions *Phys. Rev.* B **4** 592
[48] Liu Y, Demirbas U, Kellert M, Thesinga J, Cankaya H, Hua Y, Zapata L E, Pergament M and Kärtner F X 2020 Eight-pass Yb:YLF cryogenic amplifier generating 305-mJ pulses *OSA Continuum* **3** 2722–29
[49] Giesen A, Hugel H, Voss A, Wittig K, Brauch U and Opower H 1994 Scalable concept for diode-pumped high-power solid-state lasers *Appl. Phys.* B **58** 365–72
[50] Giesen A and Speiser J 2007 Fifteen years of work on thin-disk lasers: results and scaling laws *IEEE J. Sel. Top. Quantum Electron.* **13** 598–609
[51] Saraceno C J, Sutter D, Metzger T and Ahmed M A 2019 The amazing progress of high-power ultrafast thin-disk lasers *J. Eur. Opt. Soc. Rapid Publ.* **15** 7
[52] Jung R, Tummler J and Will I 2016 Regenerative thin-disk amplifier for 300 mJ pulse energy *Opt. Express* **24** 883–87
[53] Reagan B A, Curtis A H, Wernsing K A, Furch F J, Luther B M and Rocca J J 2012 Development of high energy diode-pumped thick-disk Yb:YAG chirped-pulse-amplification lasers *IEEE J. Quantum Electron.* **48** 827–35

[54] Reagan B A *et al* 2018 Scaling diode-pumped, high energy picosecond lasers to kilowatt average powers *High Power Laser Sci. Eng.* **6** e11
[55] Chi H, Dehne K A, Baumgarten C M, Wang H C, Yin L, Reagan B A and Rocca J J 2018 In situ 3-D temperature mapping of high average power cryogenic laser amplifiers *Opt. Express* **26** 5240–52
[56] Gronloh B, Russbueldt P, Schneider W, Jungbluth B and Hoffmann H D 2012 High average power sub-picosecond pulse generation at 515 nm by extracavity frequency doubling of a mode-locked Innoslab MOPA *Conf. on Solid State Lasers XXI—Technology and Devices* (San Francisco, CA: SPIE—The International Society for Optical Engineering)
[57] Rocker C, Loescher A, Bienert F, Villeval P, Lupinski D, Bauer D, Killi A, Graf T and Ahmed M A 2020 Ultrafast green thin-disk laser exceeding 1.4 kW of average power *Opt. Lett.* **45** 5522–25
[58] Phillips J P *et al* 2016 High energy, high repetition rate, second harmonic generation in large aperture DKDP, YCOB, and LBO crystals *Opt. Express* **24** 19682–94
[59] Bayramian A J *et al* 2009 High average power petawatt laser pumped by the Mercury laser for fusion materials engineering *Fusion Sci. Technol.* **56** 295–300
[60] Chi H, Wang Y, Davenport A, Menoni C S and Rocca J J 2020 Demonstration of a kilowatt average power, 1 J, green laser *Opt. Lett.* **45** 6803–6
[61] Fan G, Carpeggiani P A, Tao Z, Coccia G, Safaei R, Kaksis E, Pugzlys A, Légaré F, Schmidt B E and Baltuška A 2021 70 mJ nonlinear compression and scaling route for an Yb amplifier using large-core hollow fibers *Opt. Lett.* **46** 896–99

Chapter 2

High-performance, thulium-doped, ultrafast fiber lasers

Christian Gaida, Martin Gebhardt, Cesar Jauregui and Jens Limpert

2.1 Introduction

Since the first demonstration of the laser in 1960 [1], the field of laser physics has undergone an unparalleled development that is, still today, driven by the ever-growing demands of applications in industry as well as in fundamental and applied research. In this context it should be stressed that, besides the unique properties of laser light, such as coherence, directionality, monochromaticity and high intensity, many applications also benefit from mode-locked laser operation. This unique laser operation regime can provide broadband emission and intense ultrafast optical pulses from tabletop setups. Thus, since their first demonstration, ultrafast lasers have become indispensable tools for highly precise metrology and high-resolution spectroscopy and they keep on enabling further technological advances and innovations of relevance for our everyday life. Benefiting from the chirped pulse amplification (CPA) scheme [2], the peak-power scaling of affordable, tabletop, ultrafast lasers with diffraction-limited beam quality has made high intensities ($>10^{14}$ W cm^{-2}) available in many scientific laboratories worldwide, revolutionizing research possibilities.

The long-standing workhorse for the generation of intense laser pulses has been Ti:Sa CPA technology operating at a center wavelength of 800 nm. However, thermal effects limit the average power of this technology to just a few tens of Watts. This is a serious hindrance for many applications since, alongside the quest for high intensities, many of them can greatly benefit from a high pulse repetition rate, i.e., high average power, since it would significantly reduce the data-acquisition time and enhance the signal-to-noise ratio. To address this demand, high average power, ultrafast laser concepts have been developed over the last decades, most prominently the thin-disk laser [3, 4], the slab laser [5] and the fiber laser [6, 7]. All of them reduce thermo-optical issues due to the geometry of their active medium and are

doi:10.1088/978-0-7503-2536-3ch2

predominately operated at 1 μm wavelength. In fact, ytterbium (Yb^{3+}), with an emission band around 1 μm, is currently the most prominent rare-earth dopant for high-power lasers. The reason lies in the unique properties of Yb^{3+} in contrast to nearly all other rare-earth ions, which include a simple energy-level structure, low quantum defect and high quantum efficiency.

Among the novel high average power, ultrafast laser architectures, the fiber laser benefits from its inherent compactness and robustness, its long-term stability and its high efficiency, which is a direct result of the confinement of both the laser and the pump radiation in a waveguide. Unfortunately, this tight confinement of the laser radiation in the fiber core together with the long interaction length is also the Achilles' heel of this technology since it promotes the onset of nonlinear effects.

At this point, it should be stressed that the performance of a laser source defines the range of possible applications for that radiation and the quality of the achieved effect. Performance is a very broad term that encompasses widely different parameters such as stability, beam quality, peak and average power, pulse energy... but also, crucially, the central emission wavelength of the laser. This last parameter is extremely important since it outright grants or denies access to applications. Thus, a significant shift of the emission wavelength can open up entirely new fields of applications. For example, the increased water absorption at, e.g., around 1.9 μm, allows for a variety of biological and medical applications [8, 9] such as, e.g., tumor resection, meniscectomy or spine and eye surgery. Furthermore, an enhanced absorption at longer wavelengths can also be observed for many non-metal solids, such as polymers [10], and this can potentially lead to improvements in materials processing. Moreover, these longer wavelengths are very beneficial as a springboard for the generation of ultrafast mid-infrared (mid-IR) sources [11], which are in high demand. In fact, in recent years there has been a substantial increase in applications demanding lasers that operate at significantly longer wavelengths than the 1 μm provided by Yb^{3+}-doped materials. A common strategy to obtain the desired long-wavelength laser radiation is based on optical-parametric amplification (OPA) schemes incorporating nonlinear crystals that are pumped by laser systems at 1 μm wavelength [12]. However, there are serious obstacles to the power scaling of OPAs imposed by their inherently low quantum efficiency and thermal limitations in the nonlinear crystals. For example, a >1 kW average power, ultrafast pump laser system at 1 μm wavelength is required in order to obtain 100 W class laser radiation in the 2 μm wavelength region with an OPA [13].

There are also serious challenges to overcome when shifting the direct emission wavelength of fiber laser systems. One of these is the fact that none of the alternative laser-active dopants can reach ytterbium-doped silica fibers in terms of efficiency in practical setups. One potential exception to that is thulium (Tm^{3+}). For example, thulium-doped silica fibers [14] emitting around 2 μm would indeed offer an absorption band at around 1600 nm resulting in low quantum defect operation and, hence, in high efficiency, but high-power diode lasers are simply not available at that wavelength. This problem could be overcome by employing long-wavelength Erbium-fiber lasers or Raman lasers as pump sources, but this approach is still in its infancy. Therefore, high-power, direct diode-pumping of thulium-doped fiber lasers is currently only practical at 793 nm wavelength. Crucially, the high quantum defect between the attractive 793 nm diode laser pumping and the 2 μm emission can be partly circumvented by a so-called 2-for-1 excitation process [15], which can potentially lead to a slope efficiency of ~80%. However, in practice, even record demonstrations in terms of efficiency in high-power Tm^{3+}-doped fiber laser systems are in the range of 60% [16], and, therewith, significantly below ytterbium standards with typically >90% efficiency. In fact, ultrafast thulium-doped fiber laser systems were considered a laboratory curiosity just a few years ago. This was because the emitted peak and average-power levels were roughly two orders of magnitude lower than those achieved by state-of-the-art ytterbium-doped femtosecond fiber lasers at 1 μm wavelength. At that point, the difficulties of power scaling were widely acknowledged: the severe challenges associated with the complex energy-level structure and the lower efficiency of thulium, the lack of high-power pump diodes and optical components as well as the atmospheric water-vapor absorption lines that overlap with the gain bandwidth of thulium [17].

The last point mentioned above highlights the fact that, on top of the active material considerations, fundamental light properties have to be taken into account when shifting the laser emission toward longer wavelengths. In this context, an important quantity of a laser beam is the beam parameter product BPP which, for a given M^2-value, scales linearly with the wavelength. Consequently, the beam waist becomes linearly larger for longer wavelengths for any given fixed divergence (i.e., given optics). Hence, to achieve an application-relevant peak intensity at the target, a longer wavelength will require a higher peak power. In addition, water-vapor absorption in ambient air is responsible for an additional challenge of high-power laser beam propagation at (some) longer wavelengths. If a spectral overlap with those strong absorption lines exists, then the resulting thermal lens will lead to blooming effects, which spoil the spatial coherence of the laser emission [17].

Despite these challenges, longer operation wavelengths are still very attractive, since they conceal a substantial power-scaling potential. This is because most common limitations in fiber lasers, such as nonlinearities [18] or transverse mode instability [19], scale favorably with the wavelength. Furthermore, thulium-doped fused silica fibers have been steadily attracting attention in recent years since their benefits for high-power, ultrafast pulse operation at long wavelengths can potentially outweigh the challenges associated with this technology. In fact, thulium-doped fibers provide a very broad gain bandwidth that theoretically supports the direct amplification of <70 fs pulses. To underpin this, the following sections summarize these scaling opportunities and present, exemplarily, recent demonstrations of high peak and high average power laser systems, which are on par or even outperform their ytterbium-doped fiber counterparts.

2.2 Fundamentals, challenges and hidden scaling potential of ultrafast thulium-doped fiber lasers

In this chapter, the excitation process in thulium-doped silica fibers and challenges related to propagation effects of the 2 μm radiation in ambient air are summarized.

This is followed by a discussion of the concealed scaling potential of ultrafast thulium-doped fiber lasers compared to their mature ytterbium-doped counterparts.

2.2.1 Thulium-doped fused silica as a gain material

In contrast to the trivalent ytterbium ion (Yb^{3+}), which is the most commonly used ion for high-power laser amplification, the trivalent thulium ion (Tm^{3+}) has a very complex energy-level structure. On the one hand, this can be regarded as an advantage, since there are many possible routes for the creation of population inversion. However, on the other hand, this complexity also allows for various unwanted effects that can significantly reduce the efficiency of the laser, such as excited-state absorption, cascaded up-conversion processes and energy-transfer events between neighboring ions. To illustrate this, figure 2.1 shows a simplified energy level diagram of trivalent thulium ions [20, 21].

Figure 2.2 shows the effective cross-sections for absorption and emission based on the spectroscopic data of Tm^{3+}-doped fused silica. The emission band of interest in the wavelength range 1750–2100 nm originates from the $^3F_4$ manifold.

In order to target this emission band, several different schemes with optical pumps around 790 nm, 1200 nm and in the range between 1550 nm and 1900 nm are

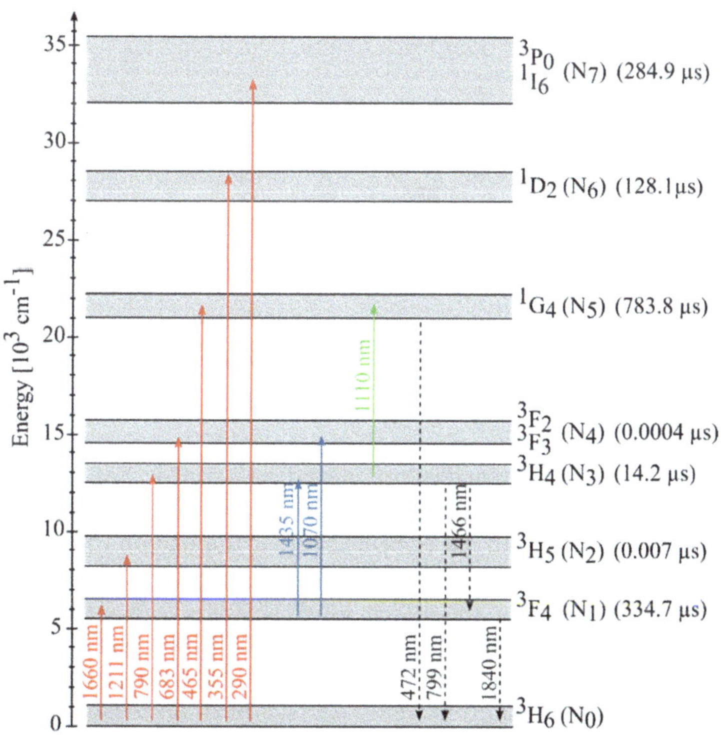

Figure 2.1. Energy level diagram of trivalent thulium ions doped in fused silica; red: ground-state absorption, blue and green: excited-state absorption, black: emission lines.

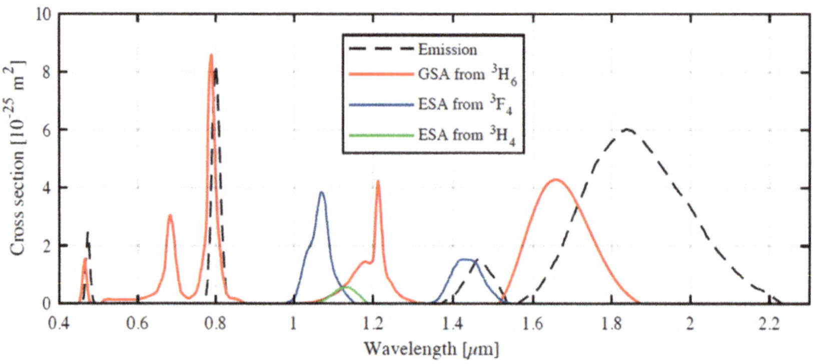

Figure 2.2. Effective cross-sections of Tm-doped fused silica glass for emission (dashed line), ground-state absorption (GSA—red line) and excited-state absorption (ESA—blue and green lines) at 298 K. Adapted with permission of Springer from [21] (2004).

possible and have been investigated [22–25]. Although all these pumping schemes nowadays allow for >10 W average output power, currently the most popular pump wavelength is 790 nm, because high-power pump diodes are readily available allowing for relatively simple high-power laser setups. One obvious major drawback of pumping at 790 nm is the large quantum defect, which can be nearly 60% for a signal wavelength of 1960 nm. This leads to a high heat load upon amplification with all its unwanted consequences.

Fortunately, a beneficial cross-relaxation process, which can significantly increase the efficiency and thereby reduce the quantum defect heating of the gain material, can be exploited when pumping at 793 nm. As shown in figure 2.3, an excited Tm^{3+}-ion is first excited by a single pump photon at 793 nm to the pump level ($^3H_4$). It then transfers part of its energy to another, neighboring, Tm^{3+}-ion in the ground state, so that both ions are eventually excited to the upper laser level ($^3F_4$). However, an efficient cross-relaxation process relies on a high doping level of, at least, ~1.5 wt% of trivalent thulium ions, since high concentrations reduce the active ion-to-ion distance in the host material, thus facilitating energy transfer among them. If this process occurs in a significant portion of the Tm^{3+}-ions, the slope efficiency of the amplification system can surpass the Stokes limit. Figure 2.4 illustrates this by plotting the published slope efficiency data of Tm^{3+}-doped fiber laser systems as a function of the doping concentration. Indeed, it can be clearly seen in figure 2.4 that using high doping concentrations is beneficial for the exploitation of this 2-for-1 process leading to slope efficiencies in Tm^{3+}-doped fiber laser systems as high as ~75%, i.e., significantly above the Stokes limit [15, 26].

However, it is worth emphasizing at this point that the experimental conditions strongly influence the contribution of cross-relaxations to the amplification process and, therefore, the efficiency of a Tm^{3+}-doped fiber laser system. In order to understand this, two additional effects need to be considered: (A) the influence of the upper laser and pump level population on the cross-relaxation rate and (B) the temperature dependence of the emission cross-sections. The first of these effects, i.e.,

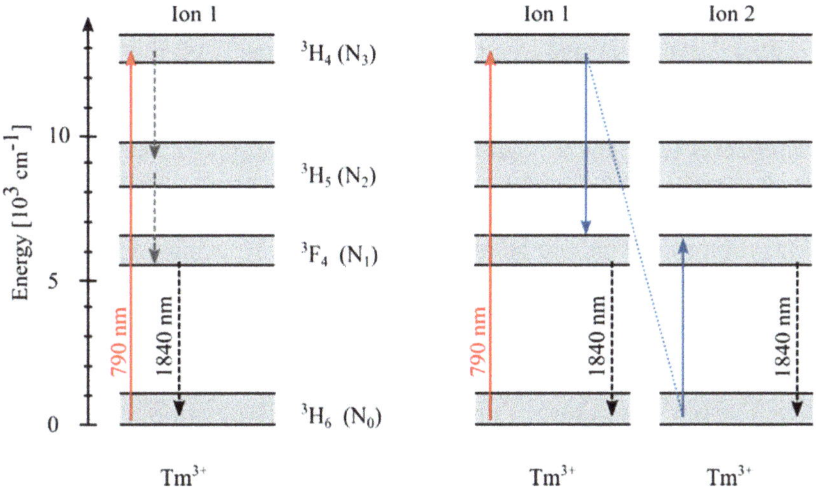

Figure 2.3. (left) Optical pumping at 793 nm and (right) optical pumping at 793 nm supported by a cross-relaxation process (blue arrows) known as 2-for-1 pumping.

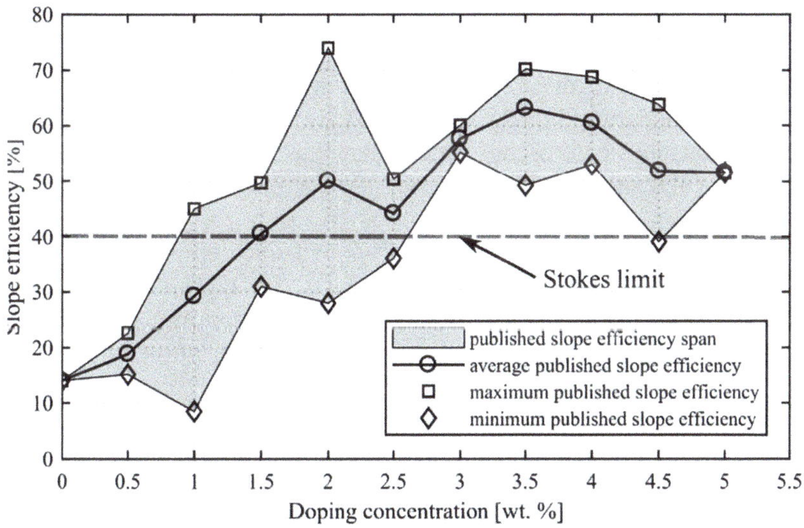

Figure 2.4. Evolution of the published slope efficiency for different ion doping concentrations in Tm^{3+}-doped fiber lasers. The upper and lower boundaries of the shaded region (squares and diamonds) show the maximum and minimum published slope efficiency for a given Tm^{3+}-doping concentration, respectively. The solid black line with circles shows the average published slope efficiency for a given Tm^{3+}-doping concentration. The data has been extracted from [16] and references therein.

the impact of the population of the upper laser and pump levels, can be evaluated by looking at the cross-relaxation rate, which can be written as [20]:

$$CR = c_{3101}N_0N_3 - c_{1310}N_1^2 \tag{2.1}$$

with N_0 being the population density of the ground state, N_1 the population density of the upper laser level and N_3 the population density of the pump band (see figure 2.3). Additionally, c_{3101} and c_{1310} are transition probabilities for the cross-relaxation processes. As revealed by equation (2.1), a high population density of the upper laser level (N_1) has a negative impact on the cross-relaxation rate. This is because, as a matter of fact, a reverse cross-relaxation process also exists, which depopulates the upper laser level. Hence, a low inversion is required to obtain a high cross-relaxation rate. This can be obtained, for example, using a high seed power in the Tm^{3+}-fiber amplifier, i.e., increasing the saturation level of the amplifier.

Regarding the second effect, figure 2.5 shows the temperature dependence of emission cross-sections of Tm^{3+}-doped silica. Figure 2.5 has been calculated using the McCumber relations [27]. As can be clearly seen, the higher the temperature of the active material, the lower the peak value of the cross-section. Hence, in order to obtain a constant gain, a higher inversion level is required in active media subject to higher temperatures. This is in contradiction to the above-mentioned requirement of low inversion for a high rate of cross-relaxations. As a consequence, sufficient cooling of high-power Tm^{3+}-doped fiber laser systems is another precondition for their highly efficient operation.

2.2.2 Detrimental propagation effects in ambient air

Although the existence of distinct atmospheric molecular absorption bands in the short-wavelength infrared (SWIR) and mid-IR wavelength regions is the basis for many applications in spectroscopy, metrology and fundamental research, other applications or even the laser performance itself can be detrimentally affected by them.

Figure 2.6 illustrates the overlap of Tm^{3+}-doped silica emission bandwidth with strong water-vapor absorption lines. In particular, for high-power, ultrafast laser systems that contain long sections of free-space propagation, the occurrence of thermal blooming effects (induced by the water-vapor absorption lines) can severely

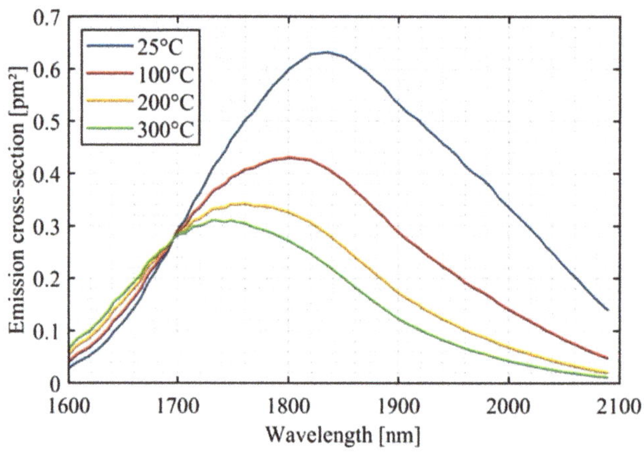

Figure 2.5. Temperature dependence of emission cross-sections of thulium-doped silica.

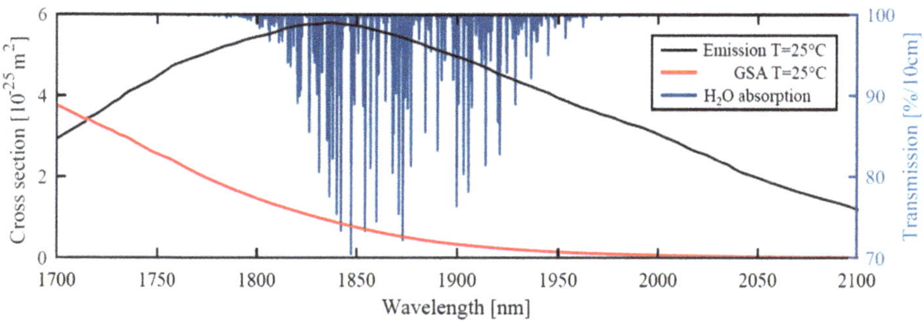

Figure 2.6. Overlap of the emission (black line) and absorption (red line) band of Tm^{3+}-doped silica with water-vapor absorption lines (blue lines). Adapted with permission from [17] © The Optical Society.

limit average-power scaling. However, even at low average powers a significant loss of pulse peak power can take place due to the dispersion introduced by the water-vapor absorption lines. This often remains unnoticed until the output of the systems since the amplification stages can easily compensate the loss in average power caused by the water-vapor absorption.

Thermal blooming arises from a thermal lens created in air by the absorption of the propagating light. This thermal lens leads to a distortion and expansion of the beam, since the thermo-optic coefficient of air is negative and, therefore, the thermal lens has a defocusing effect. Thermal blooming has been extensively studied for high-power, long-distance propagation in atmospheric transmission windows. However, the impact of this effect on ultrafast laser systems has only been thoroughly investigated recently. Figure 2.7 illustrates an experimental investigation of thermal blooming done by sweeping a narrow-linewidth laser with 24 W of average power over the spectral absorption features of water vapor located around 1913 nm. As shown, thermal blooming manifests itself as a strong beam deformation, which occurs at spectral positions that coincide with a water-vapor absorption line, provided that the power-spectral density is sufficiently high.

Pulse quality degradation occurs in addition to the linear absorption, i.e., power loss. To investigate the additional peak-power loss, the pulses emitted by a mode-locked ultrafast fiber oscillator with a spectrum spanning from 1890 nm to 1930 nm (10 dB width) were propagated over a distance of 12 m in a climate chamber with various relative humidity levels. After this propagation the autocorrelation of the pulses was measured. Additionally, the pulse propagation was simulated by incorporating the absorption-induced complex refractive index variations that are governed by the Kramers–Kronig relations. Such absorption-induced index variations translate into sharp spectral phase distortions (i.e., dispersion), which lead to the formation of a long pulse tail that can contain a significant amount of the pulse energy, as shown in figure 2.8. The excellent agreement between simulation and experimental findings obtained with this approach (shown in figures 2.8(a)–(d)) allowed for a quantitative analysis of the total peak-power loss [17]. As it turns out, the phase/dispersion-related decrease of the temporal pulse quality (see e.g., Figure 2.8(e)) significantly reduces the pulse peak power by shifting part of the

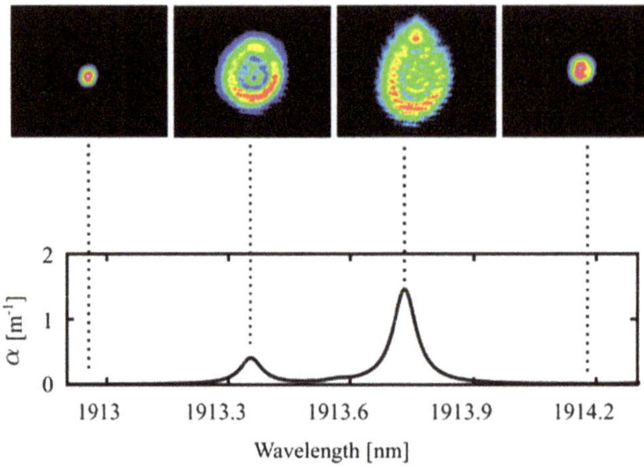

Figure 2.7. Illustration of thermal blooming upon free-space propagation of a high-power beam (upper images) as its wavelength is tuned over some water-vapor absorption lines (lower plot). Adapted with permission from [17] © The Optical Society.

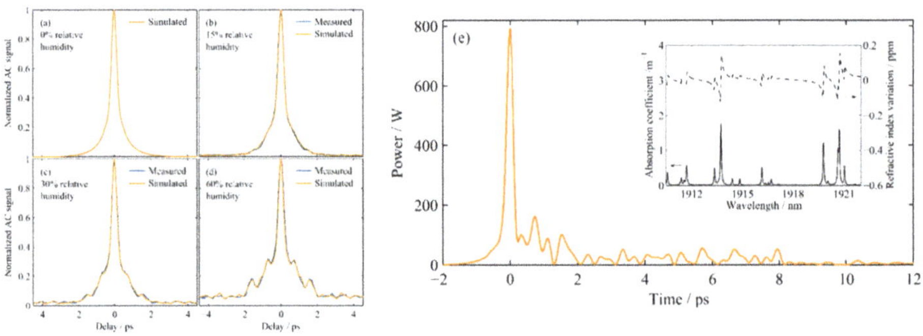

Figure 2.8. (left) Measured and simulated (a)–(d) autocorrelation traces of an originally 300 fs long pulse after 12 m of propagation in a climate chamber with different humidity levels, (right) Simulated pulse profile at 60% relative humidity level, inset: refractive index changes due to water-vapor absorption lines in a representative spectral region. Adapted with permission from [17] © The Optical Society.

energy to a long pulse tail (see figure 2.8(e)). In fact, the amount of peak power lost through this mechanism is comparable to the power fraction that is already lost due to the linear absorption term. Hence, as a rule of thumb, the peak-power loss due to sharp water-vapor absorptions is about two times higher than the loss in average power [17].

The mitigation of atmospheric molecular absorption effects is trivial for narrow-linewidth lasers, which can be tuned to operate at a wavelength that is absorption free. However, in the case of broad bandwidth, ultrafast operation, the only viable option is to completely avoid the absorption peaks. If free-space propagation is needed in the system, which is the case for most high-power ultrafast lasers since they typically feature a free-space grating-based pulse compressor, the use of dry atmosphere or, even, vacuum becomes mandatory. In principle, such a measure is not only required in the laser system, but it is often also crucial for the beam delivery path and the targeted application, as they face similar challenges.

2.2.3 The hidden performance scaling potential of 2 μm ultrafast fiber lasers

In spite of the challenges mentioned above, a transition toward longer wavelengths is an interesting opportunity for the scaling of laser performance as most properties, limitations and challenges show some kind of favorable wavelength dependence.

2.2.3.1 Gain bandwidth

A remarkable feature of Tm^{3+}-doped silica as gain material is its broad emission bandwidth, which can potentially allow for the direct amplification of sub-70 fs pulses. These pulses are roughly a factor of 2 shorter than in Yb^{3+}-doped fibers. The potential of this feature becomes even clearer when expressed in terms of optical cycles, since a 60 fs pulse at 2 μm has the same number of optical cycles as a 25 fs pulse at 800 nm. Hence, Tm^{3+}-doped fused silica offers the bandwidth prospects as 'the' broadband gain material per excellence, i.e., titanium-sapphire. This exciting fact makes Tm^{3+}-doped fused silica a very promising candidate for high peak power, ultrafast fiber amplifiers in the SWIR spectral region.

2.2.3.2 Nonlinear effects

The most common limitations in high-power fiber laser systems stem from nonlinear scattering phenomena (i.e., stimulated Raman and Brillouin scattering), which restrict the maximum output power that can be achieved in the fiber. Brillouin scattering is not observed in ultrafast amplifiers due to the broad bandwidth of the emission, but Raman can be a serious challenge. Fortunately, the Raman gain coefficient scales inversely with the wavelength, i.e., it is reduced for long-wavelength fiber lasers [18].

Even though Raman scattering is one of the first nonlinear effects to be usually observed in high-power fiber lasers, it is often not as pernicious as the nonlinear effects originating from the optical Kerr effect. As a matter of fact, self-focusing is the ultimate scaling limitation of fiber lasers since its onset usually leads to a catastrophic beam collapse which, in most cases, destroys the fiber. Crucially, even though measurements have revealed that the nonlinear refractive index n_2 is fairly constant over the entire transmission window of fused silica [28], the critical peak power for the onset of self-focusing scales with the square of the wavelength [18]. Thus, peak powers of up to 24 MW have been demonstrated for circularly polarized light at 2 μm [29]. These results significantly surpass the 6 MW limit that self-focusing imposes around 1 μm wavelength (for circularly polarized light).

Another Kerr effect, that is not as catastrophic as self-focusing, is self-phase modulation (SPM). SPM leads to a degradation of the temporal pulse quality in fiber-based chirped-pulsed amplifiers (CPA). Once again, the nonlinear phase accumulated during pulse propagation, i.e., the so-called B-integral (which is a measure for the strength of SPM), scales inverse-proportionally with the wavelength λ [18]:

$$B = \frac{2\pi n_2}{\lambda} \int_0^L \frac{\hat{P}}{A_{\text{eff}}} dz, \tag{2.2}$$

where A_{eff} is the effective mode area and $\hat{P}$ the position-dependent pulse peak power.

Another family of nonlinear effects are the parametric ones, i.e., four-wave mixing (FWM) related-effects and, in particular, modulation instabilities (MI) in the anomalous dispersion regime. MI can lead to the generation of new frequencies accompanied by pulse break-up. Despite the fact that the parametric gain also shows an inverse proportionality with the wavelength, MI can still become the most dominant nonlinear effect around 2 μm wavelength. However, the impact of FWM/MI in a 2 μm high-performance fiber laser system is still a rather unexplored topic since only few studies have been published so far [30].

2.2.3.3 Mode-area scaling potential in amplifying fibers

The guiding properties in step-index fibers are typically defined by the ratio between the core diameter and the wavelength of the guided radiation. Thus, for a given numerical aperture, the core diameter can be scaled linearly with the wavelength while preserving all relevant properties of the transverse modes. Hence, a shift of the operation wavelength from 1 μm to 2 μm should allow for two times larger core diameters, i.e., for roughly four times the mode-field area, while still preserving the single-mode condition. This property is key to mitigate most nonlinear effects (see equation (2.2)) $B = \frac{2\pi n_2}{\lambda} \int_0^L \frac{\hat{P}}{A_{\text{eff}}} dz$, and should, therefore, allow for a significant performance scaling.

The simple discussion above is, strictly speaking, only valid for index-guiding structures, such as step-index fibers. However, the general wavelength-scaling law also applies to other more advanced fiber designs.

In order to provide a hint at the mode-field scaling potential of Tm^{3+}-doped fibers, it is advantageous to have a look at the more mature Yb^{3+}-doped fiber amplifiers. There, the largest mode sizes in effective single-mode operation at 1 μm wavelength have been achieved with the large pitch fiber (LPF) concept [31]. Herewith, mode-field diameters as large as 100 μm, i.e., 100 times the wavelength of the guided light, are state-of-the-art at 1 μm wavelength. For a comparison, this figure is currently only 25 times the wavelength in the largest single-mode Tm^{3+}-doped fibers demonstrated to date. Therefore, judging by the achievements of Yb^{3+}-doped fiber amplifiers, there is still at least a factor of 4 untapped mode-field scaling potential in Tm^{3+}-doped fibers. Current research activities investigate whether advanced fiber structures such as the LPF can fully exploit the promising size scaling potential at 2 μm wavelength.

2.2.3.4 Thermal effects/transverse mode instabilities

Fiber laser systems have gained a reputation as power-scalable light sources with a power-independent excellent beam quality. Currently this reputation is threatened by a phenomenon observed for the first time about 10 years ago in 1 μm fiber laser systems, namely the effect of transverse mode instability (TMI) [19, 32, 33].

TMI occurs once the average output power reaches a threshold value and manifests itself as spatial and temporal fluctuations of the laser beam. This renders the output beam of those systems useless for many applications. Thus, TMI is currently the most limiting effect for the average-power scaling of fiber laser systems. TMI originates from a thermally-inscribed refractive index grating in the fiber core, which enables the energy transfer between the fundamental and the higher-order modes on a millisecond timescale.

At this point it should be taken into account that the mode coupling induced by a periodic index modification scales inversely with wavelength [34]. Additionally, at longer signal wavelengths the impact of thermal effects on the fiber-guiding properties is reduced (since for a given V-parameter the fiber NA is higher at longer wavelengths). Therefore, it can be expected that the mode instability threshold power is reached at significantly higher average thermal loads in 2 μm fiber laser systems than in 1 μm fiber systems.

Due to the lack of high-power, diode-based pump sources at 1600–1650 nm wavelength, the power scaling of Tm^{3+}-doped fiber laser systems has relied so far on pumping with high-power pump diodes at 793 nm. As a consequence, the quantum-defect-induced thermal load per unit fiber length is much higher than in Yb^{3+}-doped fiber laser systems at comparable average output powers. This has been regarded for a long time as a rather high burden in the context of power scaling and TMI.

However, as described above, the energy level diagram of trivalent thulium ions is, compared to that of trivalent ytterbium ions, rather complex. Thus, Tm^{3+} offers significant room for a reduction of the heat load upon amplification. For example, careful optimization of the fiber-core composition and the experimental conditions can lead to efficient cross-relaxations, as discussed previously. This approach can potentially increase the efficiency by a factor of two beyond the Stokes limit. Moreover, recent theoretical studies have unveiled that cross-relaxations might even reduce the strength of the periodic refractive index grating responsible for the onset of TMI [35]. In fact, average heat loads nearly 3 times beyond those found at the typical TMI threshold of Yb^{3+}-doped fiber laser systems have already been demonstrated [36, 37]. However, there is still a lack of experimental data and theoretical understanding of TMI at 2 μm wavelength, which is currently a subject of research in a number of laboratories around the world. In fact, a recent break-through in this context was the first observation of TMI in a kW-level Tm^{3+}-doped fiber laser system [38].

In summary, the favorable scaling of both the mode-field area and nonlinear effects indicate a significant untapped potential of Tm^{3+}-doped fiber technology with respect to the obtainable peak power. Moreover, the higher thermal resilience of long-wavelength fiber lasers holds the promise for high average powers. In addition to all of this, the gain bandwidth of Tm^{3+}-doped fused silica would support pulses as short as 70 fs. In terms of optical cycles that would correspond to a 25 fs pulse duration at 800 nm. Therefore, 2 μm silica fiber lasers might become the long-wavelength counterparts of Titanium:Sapphire lasers in the future, but in an average-power scalable platform. The fulfillment of this promise would be extremely beneficial for a large number of applications.

The following two sections present state-of-the-art, ultrafast Tm^{3+}-doped fiber laser systems, with a focus on the highest peak power and the highest average power, respectively.

2.3 High peak power femtosecond thulium-doped fiber laser

In this section, the experiment leading to the highest peak power pulses demonstrated to date with thulium-doped fiber laser systems is reviewed. A typical schematic setup of a high pulse energy Tm^{3+}-doped fiber CPA system is shown in figure 2.9. In this particular example, a commercial frequency-shifted mode-locked Erbium-fiber oscillator provides a broad spectrum in the 2 μm wavelength region at a pulse repetition rate of 25 MHz with 15 mW of average power. These seed pulses are pre-amplified in a 2.5 m long Tm^{3+}-doped photonic crystal fiber (Tm:PCF), which is end-capped, water-cooled and cladding-pumped with up to 30 W at 793 nm. This is enough pump power to reach at least 800 mW of average output power. Due to the relatively high oscillator repetition rate and the short fiber length, nonlinear effects play a negligible role at this point. Subsequently, the pulses are stretched to a full width at half-maximum (FWHM) duration of up to 1.2 ns in an Öffner-type stretcher, which has a spectral hard-cut of 1840–1980 nm. Additionally, amplitude shaping of the pulses can be realized by introducing loss at selected wavelengths of the spatially dispersed beam within the stretcher. Even though this clearly reduces the transmission of the stretcher, it is essential to counteract gain narrowing and, therewith, to harness the full potential of the gain bandwidth.

After another Tm:PCF-doped pre-amplification stage, an acousto-optical modulator (AOM) reduces the repetition rate to 781 kHz, which leaves a seed power of

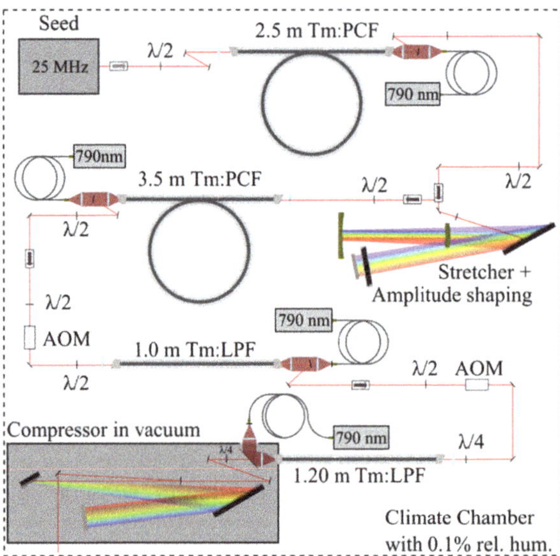

Figure 2.9. Schematic setup of a high energy fiber CPA system operating at 1.91 μm.

45 mW for the third amplification stage. This stage consists of a 1 m long Tm^{3+}-doped, LPF rod with a core diameter of 81 μm [39]. After this amplifier, the average power reaches up to 3 W. A second AOM further reduces the pulse repetition rate to 98 kHz providing a seed power of 150 mW for the final amplification stage, which is also a 1.3 m long Tm^{3+}-doped LPF rod (Tm:LPF) with a core diameter of 81 μm. The fiber is pumped with a diode laser providing up to 220 W at 793 nm. A cross-section of that fiber is shown in figure 2.10. The fused silica LPF is a very successful fiber structure design at 1 μm wavelength as it offers the largest mode areas with effective single-transverse mode operation and high-power levels. The fiber structure itself can be directly used in the 2 μm wavelength since the LPF working principle (namely the delocalization of higher order modes) works over a very broad spectral range.

A quarter-wave plate in front of the main amplifier is used to achieve circular polarization, which effectively reduces its nonlinear refractive index by 33% [40]. A second quarter-wave plate re-establishes linear polarization after the main amplifier before a Treacy-type compressor de-chirps the amplified pulses with an efficiency of >82%. Such a high-compression efficiency is enabled by dielectric gratings with reflection efficiencies of >95%, which are fabricated by the Fraunhofer IOF Jena. Figure 2.11 shows one of the compressor gratings used in the setup. The entire compressor and, with that, the entire high-power free-space propagation path, is placed in vacuum. Without that measure, neither the pulse quality nor the excellent spatial quality of the emission would have been possible, as discussed in section 2.2.2. It is for these same reasons that the entire laser is enclosed in a climate chamber held at 0.1% relative humidity.

A challenge associated with broadband operation is pulse shaping due to the aforementioned gain characteristics of the amplifiers. Without pre-compensation, the amplified stretched spectrum would generally be reduced in bandwidth (i.e., the stretched pulse duration would decrease), which can ultimately limit the extracted pulse energy in a CPA system. Hence, the goal of an optimized pre-compensation of

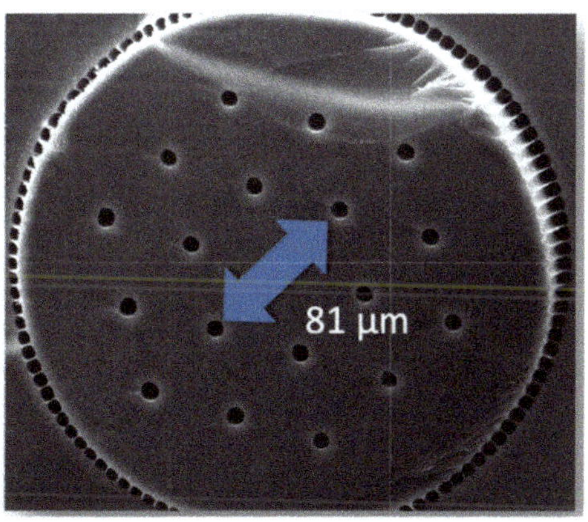

Figure 2.10. Tm^{3+}-doped large pitch fiber.

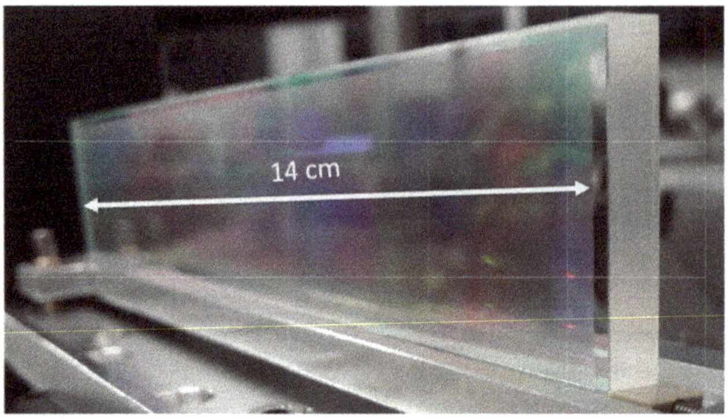

Figure 2.11. High-efficiency dielectric diffraction grating used for pulse compression at 1.9 μm wavelength.

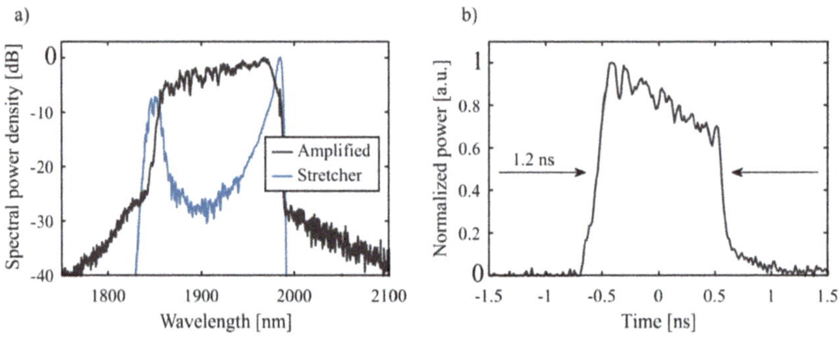

Figure 2.12. Spectral (a) and temporal (b) output characteristics of the pulse after the stretcher (blue) and after the main amplifier (black) using amplitude shaping.

the gain characteristics is a reduction of the peak power at the end of the main amplifier to minimize the impact of nonlinearities.

As illustrated in figure 2.12, the higher gain at wavelengths around 1900 nm is pre-compensated to prevent a significant temporal pulse shortening upon amplification. The goal is to keep the ~1 ns FWHM pulse duration, limited by the stretcher grating size, at the output of the system. Figure 2.12(a) compares the spectrum at the beginning of the amplifier chain with a typical amplified output spectrum. The spectral-gain landscape of the system is compensated in a way such that the final bandwidth is optimized. Consequently, the corresponding amplified stretched pulse duration could be preserved at >1 ns (see figure 2.12(b)). This is, as discussed above, an important ingredient for the optimization of the output peak power.

At a pump power of 220 W the uncompressed output from the main amplifier is 52 W. The accumulated nonlinear phase-shift (B-integral) is estimated to be about 2 rad, which is still acceptable and leads to no significant pulse distortions in the compressed output, as will be shown in the following. After the grating compressor, a compressed output power of 46 W is measured, which corresponds to a pulse energy of 470 μJ.

The retrieved temporal power envelope of the compressed pulses is shown in figure 2.13. The pulse retrieval algorithm uses the measured compressed spectrum and adds phase terms that correspond to the B-integral in the main amplifier and to a slight detuning of the stretcher-compressor pair. After that, an excellent agreement between the numerical data and the measured autocorrelation trace is obtained. Using the optimized pre-shaping and compressor settings, the FWHM duration is as short as 100 fs. This duration corresponds to only 16 optical cycles at a central wavelength of 1910 nm. The generated peak power at a repetition rate of 98 kHz is just below 4 GW.

In order to discuss the peak-power scalability of ultrafast Tm^{3+}-doped fiber laser systems, a comparison to the mature 1 μm Yb^{3+}-doped fiber technology is advisable to provide a reference for this discussion.

As described, the employed main-amplifier fiber in the system described above has a core diameter of 80 μm. In contrast, as already mentioned before, effective single-mode operation with Yb^{3+}-doped fiber amplifiers has been demonstrated with a core dimension as large as 100 μm. Hence, in theory, and assuming a technological advance in the manufacturing of Tm^{3+}-doped fibers, core sizes in the range of 200 μm (taking into account the wavelength-scaling law of the core size) should be feasible. This would represent a scaling factor for the core area of roughly a factor of 6.

The stretched pulse duration of the described experiment is, in comparison to record-performance fiber CPA systems at 1 μm, rather short, with roughly 1 ns as opposed to 4 ns in state-of-the-art Yb^{3+}-doped fiber laser systems. Since there is no fundamental reason why the dielectric compressor gratings should not be available with larger dimensions, it is reasonable to expect stretched pulse durations of 4 ns and beyond also in ultrafast Tm^{3+}-doped fiber laser systems in the future.

Overall, considering these two approaches (larger fiber cores and larger grating dimensions), a peak power as high as 100 GW appears feasible from a single aperture Tm^{3+}-doped, ultrafast fiber laser system. To stress the true significance of this prospect, it is worth pointing out that even optimized femtosecond fiber laser

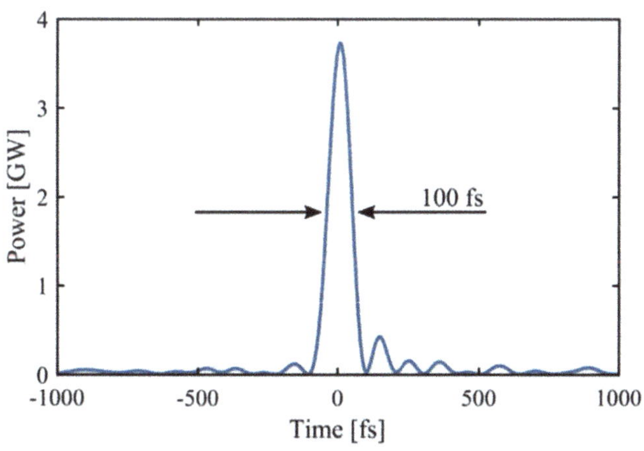

Figure 2.13. Retrieved temporal power evolution of the compressed pulse at the highest pulse energy.

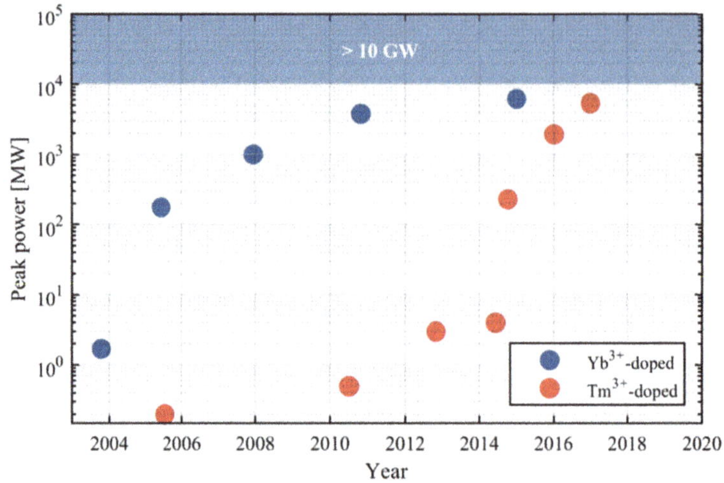

Figure 2.14. Evolution of the peak-power scaling of single-emitter, Yb^{3+}-doped (blue points) and Tm^{3+}-doped (red points) fiber laser systems over the years. Data extracted from [41–48].

systems at 1 μm wavelength struggle to reach 10 GW of peak power, as can be inferred from figure 2.14.

In summary, it can be observed in figure 2.14 that single-emitter (i.e., non-coherently combined) Tm^{3+}-doped fiber laser systems have already matched the peak-power performance of their Yb^{3+}-doped counterparts, but with the difference that their scaling potential is not saturating yet. Therefore, it can be realistically expected that Tm^{3+}-doped fiber laser systems will significantly outperform Yb^{3+}-doped fiber systems in terms of peak power in the near future.

2.4 High average power, ultrafast thulium-doped fiber lasers

Increasing the average power of Tm^{3+}-doped fibers relies, to a large extent, on the understanding of efficiency, i.e., cross-relaxations, and the associated thermal load created in the active medium. As revealed in section 2.2.1, the rate of cross-relaxation is determined by the doping concentration as well as by the experimental conditions, namely the saturation level and the fiber temperature. To achieve high efficiency and, therefore, to enable scaling the average power, both a sufficiently high cooling and a sufficiently high seed power are essential preconditions.

The experimental setup with which the highest average power demonstrated to date in an ultrafast, Tm^{3+}-doped fiber laser system [37] has been achieved is depicted in figure 2.15. This setup follows the usual architecture of a Tm^{3+}-doped fiber CPA system. The system is seeded by a commercial oscillator that delivers ultrashort pulses at 80 MHz repetition rate and at a central wavelength of 1960 nm. The pulses from the oscillator are stretched to 1 ns FWHM duration in a fiber stretcher made with chirped fiber Bragg gratings. After the stretcher, there is a remaining average power of 0.5 mW, which is then amplified to 400 mW using a 2 m long, double-clad, polarization-maintaining, Tm^{3+}-doped fiber. This active fiber is cladding-pumped in

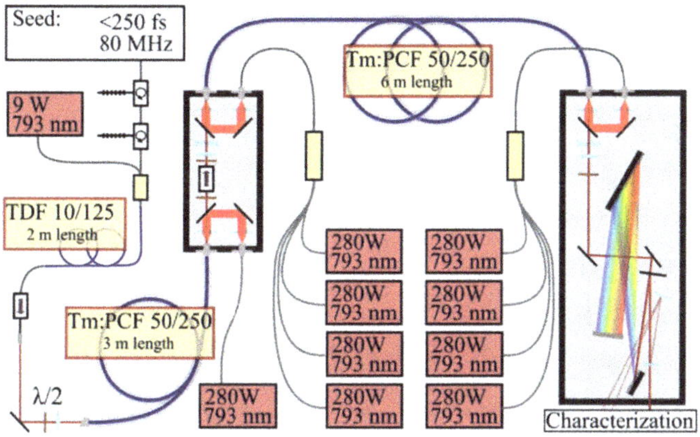

Figure 2.15. Setup of a kW average-power, ultrafast Tm^{3+}-doped fiber CPA system. Adapted with permission from [37] © The Optical Society.

the co-propagating direction at 793 nm. The amplified signal, which has an average power of 300 mW, is free-space coupled to a second amplification stage after going through an isolator. The second amplifier is built using a 3 m long Tm^{3+}-doped, photonic crystal fiber (Tm:PCF) which has a core/cladding diameter of 50 μm/ 250 μm. The doping concentration of this fiber is 7 × 1025 ions/m^3. The Tm:PCF of the second amplifier is cladding-pumped in the counter-propagating direction at 793 nm wavelength with up to 280 W. Even though this would be enough to amplify the signal to >100 W of average power, the seed for the main amplifier has been limited to 10 W to prevent the creation of a thermal lens in the free-space isolator located in front of the final amplifier. The high-power main amplifier also uses the Tm:PCF but, this time, it is 6 m long because it is bidirectionally pumped with eight pump modules, each delivering 280 W of CW power at 793 nm. The pump modules have been efficiently combined using two 4:1 combiners (one at each end of the main amplifier fiber) so that a total pump power of 2090 W is available. Both the fiber of the main-amplifier and the final section of the pump delivery fiber are water-cooled to avoid any thermal damage. The pump light is coupled into the Tm:PCF main amplifier using an arrangement of specially designed mirrors and lenses.

After the main amplifier, the pulses are re-compressed in a Treacy-type compressor, which contains very efficient dielectric reflection gratings, something that was already mentioned in section 3. The high-power section of the system is enclosed in a vacuum chamber filled with 1 bar of dry helium gas to avoid detrimental water-vapor absorption effects [17], as discussed in section 2.2. A beam sample is taken at the output of the compressor using the reflections from a fused-silica wedge to measure the temporal, spectral, and spatial characteristics of the high-power output.

Using the system described above, it is possible to reach a pump-limited, maximum average output power of 1150 W at the output of the main amplifier fiber, as illustrated in figure 2.16. Hereby, if a free-space pump-launching efficiency of 90% is assumed, it is possible to determine that the slope efficiency of this system is as high as 61%. This high slope efficiency can only be reached if there is a strong

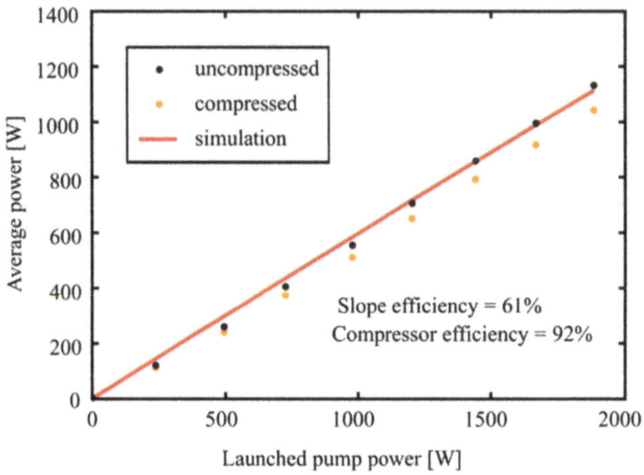

Figure 2.16. Measured output average power at a 1960 nm center wavelength before (black) and after (orange) the compressor as a function of the launched pump power at 793 nm. Adapted with permission from [37] © The Optical Society.

contribution of cross-relaxations in the high power Tm:PCF (figure 2.16). Such high slope efficiency has been achieved by following the guidelines outlined in section 2.1, namely sufficient cooling and seeding.

The amplification process of this system has been modelled by solving the steady-state rate equations of Tm^{3+}-doped fused silica fibers. These equations include the effect of cross-relaxations, energy transfer up-conversion processes, as well as the temperature dependence of the emission cross-section. The simulation results can closely reproduce the measured power slope, as shown in figure 2.16. Additionally, since the steady-state population and transition rates of each energy level are known at each position along the fiber, it is possible to obtain the total number of non-radiative transitions. Therefore, using this information, it is possible to calculate both the heat-load and the transversal temperature profile along the fiber. This is shown in figure 2.17, where the heat load in the core along the active fiber is plotted at 1150 W of average output power. As can be seen, the average heat-load reaches a value of 98 W/m, which is significantly higher than the typical TMI threshold in Yb^{3+}-doped fiber laser systems at 1 μm wavelength (~30 W/m).

The compressor efficiency has been determined to be 92% at all power levels by comparing the power slope before and after the compressor. Figure 2.18 shows the intensity- autocorrelation at the highest average output power of 1060 W, which reveals a pulse duration of 265 fs. This is only marginally longer than the transform-limited pulse duration (calculated using the measured spectrum) of 260 fs, in spite of an estimated accumulated nonlinear phase of 3 rad.

The system also exhibits an excellent beam quality with an M^2 value of <1.1 at the highest output power (figure 2.19). It is worth pointing out that the beam quality does not change with increasing average output power. Additionally, the onset of TMI has been excluded by a frequency-resolved pointing-stability measurement [49].

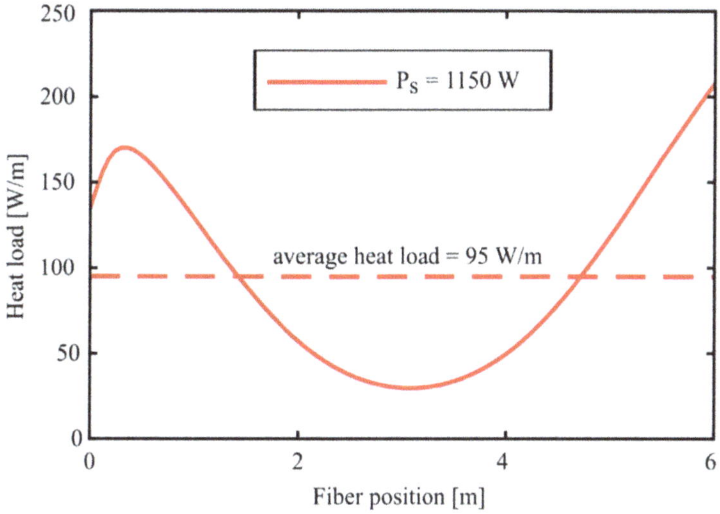

Figure 2.17. Estimated heat load in the fiber core with respect to the fiber position for an average output power of 1150 W. Adapted with permission from [37] © The Optical Society.

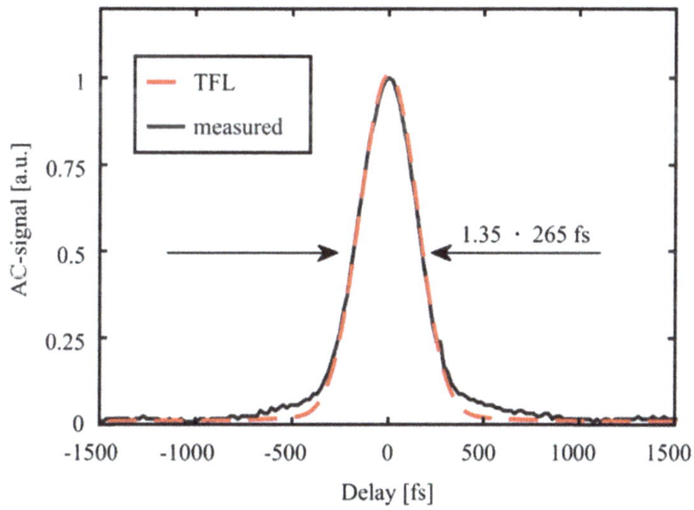

Figure 2.18. Measured intensity autocorrelation at 1060 W average output power after compression. Adapted with permission from [37] © The Optical Society.

The extracted 1150 W of average power from the main amplification stage is the highest average power reported to date from a Tm^{3+}-doped fiber laser system. At this point it is worth highlighting that the employed fiber has a very large core of 50 μm and is only 6 m long, which is extremely beneficial for mitigating nonlinear effects. In fact, there is no Yb^{3+}-doped fiber, with a similar core diameter and fiber length, capable of matching the performance of this system while providing stable diffraction-limited beam quality. Hence, this is the first experimental proof

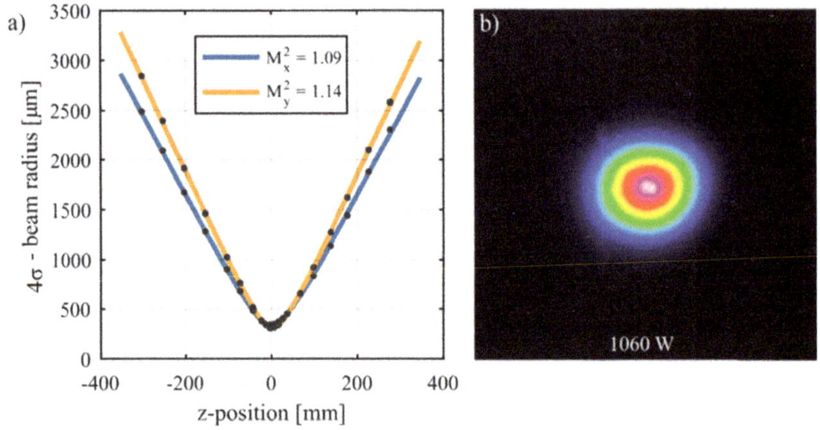

Figure 2.19. (a) M^2 measurement and (b) output beam profile at the highest average output power after pulse compression. Adapted with permission from [37] © The Optical Society.

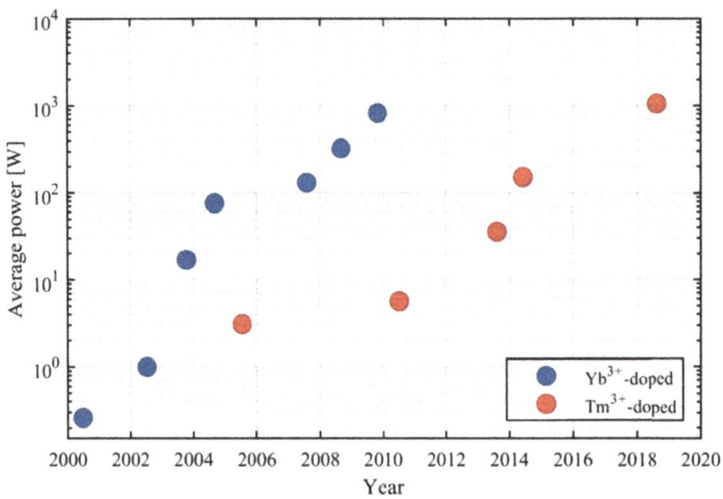

Figure 2.20. Evolution of the average-power scaling of single-emitter, Yb^{3+}-doped (blue points) and Tm^{3+}-doped (red points) fiber laser systems over the years. Data extracted from [37, 41, 42, 44, 50–54].

that a Tm^{3+}-doped fiber laser system can outperform a Yb^{3+}-doped fiber system in terms of diffraction-limited average output power, as seen in figure 2.20.

2.5 Future prospects of frequency conversion driven by high-performance, ultrafast fiber laser systems at 2 μm wavelength

As mentioned at the beginning of this chapter, a number of applications would strongly benefit from high-performance, long-wavelength, ultrafast lasers. For example, the increased water absorption, e.g., around 1.9 μm, enables a number of biological and medical applications, such as tumor resection, meniscectomy or spine and eye surgery. Furthermore, an enhanced absorption at longer wavelengths can also be observed for many non-metal solids, such as polymers, and this can potentially lead to improvements in materials processing.

In addition, there are very interesting wavelength-scaling laws for frequency conversion in application-relevant spectral regions, namely the soft x-rays, mid-IR and THz. All of those spectral regions enable new fields of applications. For example, coherent mid-IR radiation is currently the subject of intense research, driven by the substantial number of existing and emerging applications in medical diagnostics, spectroscopy, microscopy and fundamental science [55]. On the other hand, terahertz imaging and spectroscopy already enable nondestructive chemical analysis [56–58]. Its applications run from detecting defects in industrial and pharmaceutical product inspection, to spectroscopy, as well as material characterization, the detection of concealed weapons at airports or the detection of cancer and caries. Finally, coherent soft x-ray radiation, in particular in the water-window spectral region, is an enabling tool for time-resolved spectroscopy and imaging of biological samples. Hence, it allows investigating relevant structures in life-science with a very high temporal and spatial resolution [59].

As in most applications, the flux of the driving source determines the quality of the application itself. In the following, the wavelength-scaling laws are summarized to highlight the potential of an ultrafast Tm^{3+}-doped fiber laser system as a driver of frequency-conversion processes.

2.5.1 High harmonic generation (HHG) into the water window

Water-window, soft x-ray radiation between the K-absorption edge of carbon at a photon energy of 282 eV (4.4 nm) and the K-absorption edge of oxygen at 533 eV (2.33 nm) penetrates water while nitrogen and other elements found in biological specimens still absorb it [59]. Hence, this radiation is well-suited for spectroscopic and microscopic investigations on biological (even living) specimens in their natural environment. The flux of the illumination source determines the achievable resolution, penetration depth and/or imaging rate. Therefore, a high-power, tabletop, coherent soft x-ray source can be considered as an enabling tool for imaging techniques such as XUV coherence tomography (XCT) [60], lens-less coherent diffractive imaging (CDI) [61] as well as high-resolution, Fourier-transform and absorption spectroscopy. These techniques will pave the way for new approaches in biological and medical investigations, for example to detect and analyze the tiniest protein structures, which could be essential for future diagnosis and treatment of diseases. In addition to imaging and spectroscopy, which are the typical application fields of synchrotrons, the ultrashort duration of the soft x-ray pulses generated by laser-based sources allow resolving the temporal dynamics of chemical reactions [62].

The generation of spatially coherent light in the water-window region via gas harmonics is a promising but very challenging approach. Even the most intense driving femtosecond laser sources at 800 nm or around 1 μm wavelength struggle to produce an application-relevant flux above 100 eV photon energy. Hereby, an

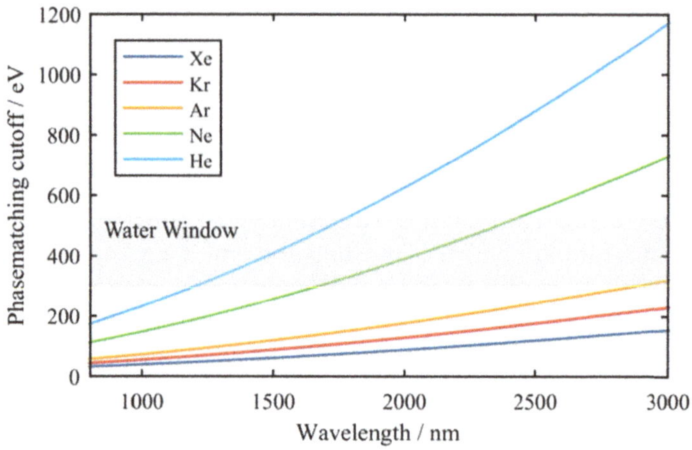

Figure 2.21. Calculated phase-matching cut-off as a function of the HHG driving laser wavelength.

increase of the driving laser wavelength is the most promising path to generate coherent radiation in this spectral region of high interest, even in spite of the fact that the reduced recombination probability drastically lowers the HHG conversion efficiency [63]. The reason is the (wavelength)^2-scaling [64] of the ponderomotive potential and, therewith, the phase-matching cut-off energy of the harmonics [63], as illustrated in figure 2.21 for different noble gases. In order to achieve application-relevant flux at 280–530 eV photon energy, the most important component is a high flux driving laser at a carefully chosen wavelength. As can be seen in figure 2.21, the lowest and, therefore, most efficient driving laser wavelength that can be used to reach the full water window via HHG employing few-cycle pulses is ~1.8 μm. High-power emission near this wavelength can be realized with a Tm^{3+}-doped fiber laser system.

2.5.2 High-power source in the fingerprint mid-IR spectral region

Studying ro-vibrational modes of organic molecules is of paramount interest in the context of modern fundamental and life-sciences. Many of these molecules feature an absorption band in the wavelength region between 2 μm and 20 μm, often referred to as the molecular fingerprint region. This motivation drives the ever-growing demand for high flux sources in this wavelength region [65–69]. Of course, synchrotrons can cover this spectral region but, their size, cost and accessibility restrict a widespread use. In this context, laser-based sources might become an enabling technology for future spectroscopic applications. However, to date, there is no laser-based source that can cover the entire molecular fingerprint region and combines the key properties desirable for coherent spectroscopy: broad wavelength coverage, spatial and temporal coherence, phase stability (frequency comb), high repetition rate and high average power. The latter is especially important since the shot-noise limited SNR increases with increasing power. Due to the lack of broadband mid-IR laser gain media, coherent broadband mid-IR sources usually employ parametric down-conversion of ultrashort laser pulses generated by matured visible/near-IR laser technologies. Driving these processes with 1 μm lasers comes with a significant restriction in the selection of nonlinear media [70]. On top of that, the conversion efficiency to long wavelengths is fundamentally limited as a result of the Manley–Rowe relations [71]. The natural choice for the conversion of radiation into the mid-IR is non-oxide crystals, which feature high transmission in the whole molecular fingerprint region and possess high nonlinear coefficients and damage thresholds. Yet these crystals feature high linear or nonlinear absorption at 1 μm wavelength. However, shifting to a 2 μm driver laser makes these non-oxide crystals very appealing [70] because they are transparent at 2 μm wavelength and, due to the lower photon energy, the detrimental impact of linear and nonlinear absorption effects is mitigated. On top of this, following the Manley–Rowe relations [71], a conversion to longer wavelengths will be more efficient using longer driving wavelengths.

2.5.3 High flux THz radiation

The lack of high-power, tabletop, ultrafast THz sources (of Watt-level and beyond) remains a challenge for many THz applications such as imaging, spectroscopy and the fundamental physics [58]. These applications ask for a higher average power THz radiation to enable faster data acquisition and more precise measurements with higher signal-to-noise ratios. Yet, these high-power sources of THz radiation emitting short pulses are mostly restricted to synchrotrons [72]. Such sources currently cover the challenging frequency window of 0.1–10 THz with very high-power levels (>10 W average power [72]), albeit with low spatial coherence. However, as already mentioned above, access to these facilities is very restricted and costly. Therefore, a Watt-level, tabletop THz source would be of utmost importance for the widespread development of many applications.

Another wavelength-scaling law supports the use of air-plasma THz generation with longer driving wavelengths. It has been shown that scaling to longer wavelengths strongly boosts the down-conversion efficiency in the THz region. A $(\text{wavelength})^{4.6}$ scaling of the THz yield has been reported, which strongly supports the use of a 2 μm driving wavelength [73]. In particular, conversion efficiencies from 1.8 μm radiation to the THz range as high as 10^{-3}, i.e., more than one order of magnitude higher than in previous plasma-based experiments performed with an 800 nm pump wavelength at a similar laser pump energy, have been observed [73]. Hence, the combination of a power-scalable, Tm^{3+}-doped, ultrafast laser technology and a favorable scaling of the conversion efficiency toward the THz region will allow generating Watt-level average powers of broadband THz radiation, i.e., developing a tabletop alternative to synchrotrons in this spectral region.

2.6 Conclusion

Currently more and more applications are demanding the development of high-power ultrafast sources at longer wavelengths than those offered by current high-power laser systems. Tm^{3+}-doped ultrafast fiber laser technology has been identified as an appealing candidate for such a development due to its high-gain, broad emission bandwidth, potential high efficiency and compatibility with the mature fused silica technology and with current high-power pump diodes. In spite of the enticing prospects for Tm^{3+}-doped ultrafast fiber laser technology, its development and, in particular, the power scaling, has taken a long time to take off. The main reason is that these systems face serious challenges when operating at high power, such as a very high thermal load, the impact of water-vapor absorption and operation in the anomalous dispersion region of the optical fibers. Even though ultrafast Tm^{3+}-doped fiber technology has significantly profited from the knowledge acquired over the years in the now mature Yb^{3+}-doped fiber systems, the previously mentioned challenges are exclusive to Tm^{3+}-doped fiber technology. Therefore, it took some time to overcome and learn to deal with them. This slow start in the power scaling led to the impression that this technology might not be well-suited for high-power operation. However, as it has been shown in this chapter, not only are Tm^{3+}-doped ultrafast fiber laser systems able to deliver high powers, but they have even already matched the performance of ultrafast Yb^{3+}-doped fiber systems.

Tm^{3+}-doped ultrafast fiber laser systems have a concealed scaling potential that has been unveiled in this chapter. This potential is grounded in the fact that the longer operating wavelength leads to a higher onset power for nonlinear effects and allows for larger fiber-core diameters (without changing the guiding properties), which further reduce the impact of nonlinear effects. Additionally, it has been shown that Tm^{3+}-doped fibers seems to be more resilient to the effect of TMI than their Yb^{3+}-doped counterparts. Moreover, the broad gain bandwidth of Tm^{3+}-doped fibers together with the longer emission wavelength are very helpful to obtain high peak power, few-cycle pulses. Therefore, Tm^{3+}-doped silica fiber lasers might become the long-wavelength counterparts of Titanium:Sapphire lasers in the future, but in an average-power scalable platform.

Taking into account all the performance-scaling advantages of this technology discussed in this chapter, a peak power >100 GW (>10 mJ pulse energy and <100 fs pulse duration) together with kW-level of average power might be feasible with Tm^{3+}-doped fiber laser systems in the future. This represents laser systems that are roughly two orders of magnitude more powerful than today's record 2 μm ultrafast fiber lasers and more than an order of magnitude above the mature 1 μm ytterbium-doped fiber laser technology. This unparalleled power-scaling potential can be further increased using coherent-beam combining techniques [74] and multicore fibers. Such exciting prospects forecast a bright future for ultrafast Tm^{3+}-doped silica fiber laser systems.

References

[1] Maiman T H 1960 Stimulated optical radiation in ruby *Nature* **187** 493–4

[2] Strickland D and Mourou G 1985 Compression of amplified chirped optical pulses *Opt. Commun.* **56** 219–21

[3] Negel J-P, Loescher A, Voss A, Bauer D, Sutter D, Killi A, Ahmed M A and Graf T 2015 Ultrafast thin-disk multipass laser amplifier delivering 1.4 kW (47 mJ, 1030 nm) average power converted to 820 W at 515 nm and 234 W at 343 nm *Opt. Express* **23** 21064
[4] Nubbemeyer T *et al* 2017 1-kW, 200-mJ, picosecond thin-disk laser system *Opt. Lett.* **42** 1381–4
[5] Russbueldt P, Mans T, Weitenberg J, Hoffmann H D and Poprawe R 2010 Compact diode-pumped 1.1 kW Yb:YAG Innoslab femtosecond amplifier *Opt. Lett.* **35** 4169–71
[6] Jauregui C, Limpert J and Tünnermann A 2013 High-power fibre lasers *Nat. Photonics* **7** 861–7
[7] Zervas M N and Codemard C A 2014 High power fiber lasers: a review *IEEE J. Sel. Top. Quantum Electron.* **20** 219–41
[8] Fried N M 2005 High-power laser vaporization of the canine prostate using a 110 W thulium fiber laser at 1.91 μm *Lasers Surg. Med.* **36** 52–6
[9] Traxer O and Keller E X 2020 Thulium fiber laser: the new player for kidney stone treatment? A comparison with Holmium:YAG laser *World J. Urol.* **38** 1883–94
[10] Mingareev I, Weirauch F, Olowinsky A, Shah L, Kadwani P and Richardson M 2012 Welding of polymers using a 2 μm thulium fiber laser *Opt. Laser Technol.* **44** 2095–9
[11] Pires H, Baudisch M, Sanchez D, Hemmer M and Biegert J 2015 Ultrashort pulse generation in the mid-IR *Prog. Quantum Electron.* **43** 1–30
[12] Shamir Y, Rothhardt J, Hädrich S, Demmler S, Tschernajew M, Limpert J and Tünnermann A 2015 High-average-power 2 μm few-cycle optical parametric chirped pulse amplifier at 100 kHz repetition rate *Opt. Lett.* **40** 5546
[13] Windeler M K R, Mecseki K, Miahnahri A, Robinson J S, Fraser J M, Fry A R and Tavella F 2019 100 W high-repetition-rate near-infrared optical parametric chirped pulse amplifier *Opt. Lett.* **44** 4287
[14] Moulton P F *et al* 2009 Tm-doped fiber lasers: Fundamentals and power scaling *IEEE J. Sel. Top. Quantum Electron.* **15** 85–92
[15] Jackson S D 2004 Cross relaxation and energy transfer upconversion processes relevant to the functioning of 2 μm Tm^{3+}-doped silica fibre lasers *Opt. Commun.* **230** 197–203
[16] Sincore A, Bradford J D, Cook J, Shah L and Richardson M C 2018 High average power thulium-doped silica fiber lasers: review of systems and concepts *IEEE J. Sel. Top. Quantum Electron.* **24** 1–8
[17] Gebhardt M, Gaida C, Stutzki F, Hädrich S, Jauregui C, Limpert J and Tünnermann A 2015 Impact of atmospheric molecular absorption on the temporal and spatial evolution of ultra-short optical pulses *Opt. Express* **23** 13776
[18] Agrawal G P 2006 *Nonlinear Fiber Optics* 3rd edn (New York: Academic)
[19] Jauregui C, Stihler C and Limpert J 2020 Transverse mode instability *Adv. Opt. Photonics* **12** 429
[20] Jackson S D and King T A 1999 Theoretical modeling of Tm-doped silica fiber lasers *J. Lightwave Technol.* **17** 948–56
[21] Peterka P, Faure B, Blanc W, Karásek M and Dussardier B 2004 Theoretical modelling of S-band thulium-doped silica fibre amplifiers *Opt. Quantum Electron.* **36** 201–12
[22] Hanna D C, Perry I R, Lincoln J R and Townsend J E 1990 A 1-Watt thulium-doped cw fibre laser operating at 2 μm *Opt. Commun.* **80** 52–6
[23] Creeden D, Johnson B R, Rines G A and Setzler S D 2014 High power resonant pumping of Tm-doped fiber amplifiers in core- and cladding-pumped configurations *Opt. Express* **22** 29067

[24] Newburgh G A, Zhang J and Dubinskii M 2017 Tm-doped fiber laser resonantly diode-cladding-pumped at 1620 nm *Laser Phys. Lett.* **14** 125101
[25] Creeden D, Johnson B R, Setzler S D and Chicklis E P 2014 Resonantly pumped Tm-doped fiber laser with 90% slope efficiency *Opt. Lett.* **39** 470
[26] Petit V, Carter A, Tumminelli R, Hemming A, Haub J and Simakov N 2018 Highly doped and highly efficient Tm doped fiber laser (conference presentation) *Fiber Lasers XV: Technology and Systems* ed A L Carter and I Hartl (Bellingham, WA: SPIE) p 21
[27] McCumber D E 1964 Einstein relations connecting broadband emission and absorption spectra *Phys. Rev.* **136** A954–7
[28] Milam D 1998 Review and assessment of measured values of the nonlinear refractive-index coefficient of fused silica *Appl. Opt.* **37** 546
[29] Gaida C, Gebhardt M, Stutzki F, Jauregui C, Limpert J and Tünnermann A 2015 Self-compression in a solid fiber to 24 MW peak power with few-cycle pulses at 2 μm wavelength *Opt. Lett.* **40** 5160
[30] Cook J, Bradford J D, Bodnar N, Richardson M C and Sincore A M 2018 Experimental investigation on varying spectral bandwidth when amplifying a pulsed superfluorescent 2-μm source in Tm:fiber *Fiber Lasers XV: Technology and Systems* ed A L Carter and I Hartl (Bellingham, WA: SPIE) p 24
[31] Limpert J, Stutzki F, Jansen F, Otto H-J, Eidam T, Jauregui C and Tünnermann A 2012 Yb-doped large-pitch fibres: effective single-mode operation based on higher-order mode delocalisation *Light: Sci. Appl.* **1** e8
[32] Eidam T, Wirth C, Jauregui C, Stutzki F, Jansen F, Otto H, Schmidt O, Schreiber T, Limpert J and Tünnermann A 2011 Experimental observations of the threshold-like onset of mode instabilities in high power fiber amplifiers *Opt. Express* **19** 13218–24
[33] Otto H-J, Stutzki F, Jansen F, Eidam T, Jauregui C, Limpert J and Tünnerman A 2012 Experimental study of mode instabilities in high power fiber amplifiers *Advanced Solid-State Photonics, ASSP 2012* (Baltimore, MA: Optical Society of America)
[34] Erdogan T 1997 Fiber grating spectra *J. Lightwave Technol.* **15** 1277–94
[35] Arlee J J S, Smith V, Smith A V and Smith J J 2015 Mode instability thresholds for Tm-doped fiber amplifiers pumped at 790 nm *Opt. Express* **2304** 11407–22
[36] Goodno G D, Book L D and Rothenberg J E 2009 Low-phase-noise, single-frequency, single-mode 608 W thulium fiber amplifier *Opt. Lett.* **34** 1204
[37] Gaida C *et al* 2018 Ultrafast thulium fiber laser system emitting more than 1 kW of average power *Opt. Lett.* **43** 5853
[38] Gaida C, Gebhardt M, Heuermann T, Wang Z, Stutzki F, Jauregui-Misas C and Limpert J 2019 Observation of transverse-mode instabilities in a thulium-doped fiber amplifier *Fiber Lasers XVI: Technology and Systems* vol 1089702 ed L Dong and A L Carter (Bellingham, WA: SPIE) p 1
[39] Jansen F, Stutzki F, Jauregui C, Limpert J and Tünnermann A 2012 High-power very large mode-area thulium-doped fiber laser *Opt. Lett.* **37** 4546
[40] Schimpf D N, Eidam T, Seise E, Hädrich S, Limpert J and Tünnermann A 2009 Circular versus linear polarization in laser-amplifiers with Kerr-nonlinearity *Opt. Express* **17** 18774
[41] Imeshev G and Fermann M E 2005 230-kW peak power femtosecond pulses from a high power tunable source based on amplification in Tm-doped fiber *Opt. Express* **13** 7424–31
[42] Haxsen F, Wandt D, Morgner U, Neumann J J and Kracht D 2010 Pulse energy of 151 nJ from ultrafast thulium-doped chirped-pulse fiber amplifier *Opt. Lett.* **35** 2991–3

[43] Sims R A, Kadwani P, Shah A S L and Richardson M 2013 1 μJ, sub-500 fs chirped pulse amplification in a Tm-doped fiber system *Opt. Lett.* **38** 121–3
[44] Wan P, Yang L-M and Liu J 2013 High pulse energy 2 μm femtosecond fiber laser *Opt. Express* **21** 1798–803
[45] Wan P, Yang L and Liu J 2013 156 micro-J ultrafast Thulium-doped fiber laser *Proc. SPIE* **1** 7–13
[46] Stutzki F, Gaida C, Gebhardt M, Jansen F, Jauregui C, Limpert J and Tünnermann A 2015 Tm-based fiber-laser system with more than 200 MW peak power *Opt. Lett.* **40** 9–12
[47] Gaida C, Gebhardt M, Stutzki F, Jauregui C, Limpert J and Tünnermann A 2016 Thulium-doped fiber chirped-pulse amplification system with 2 GW of peak power *Opt. Lett.* **41** 4130
[48] Gaida C, Gebhardt M, Stutzki F, Jauregui C, Limpert J and Tünnermann A 2017 90 fs pulses with >5 GW peak power from a high repetition rate Tm-doped fiber CPA system *Optics InfoBase Conf. Papers* vol Part F75-A
[49] Otto H, Stutzki F, Jansen F, Eidam T, Jauregui C, Limpert J and Tünnermann A 2012 Temporal dynamics of mode instabilities in high-power fiber lasers and amplifiers *Opt. Express* **20** 10180–92
[50] Sims R A, Kadwani P, Ebendorff-Heideprem H, Shah L, Monro T M and Richardson M 2013 Chirped pulse amplification in single mode Tm:fiber using a chirped Bragg grating *Appl. Phys.* B **111** 299–304
[51] Yang L-M, Wan P, Protopopov V and Liu J 2012 2 micron femtosecond fiber laser at low repetition rate and high pulse energy *Opt. Express* **20** 5683–8
[52] Wan P, Yang L-M and Liu J 2013 High power 2 μm femtosecond fiber laser *Opt. Express* **21** 21374
[53] Stutzki F *et al* 2014 152 W average power Tm-doped fiber CPA system *Opt. Lett.* **39** 4671–4
[54] Tan F, Shi H, Sun R, Wang P and Wang P 2016 1 μJ, sub-300 fs pulse generation from a compact thulium-doped chirped pulse amplifier seeded by Raman shifted erbium-doped fiber laser *Opt. Express* **24** 22461
[55] Cossel K C, Waxman E M, Finneran I A, Blake G A, Ye J and Newbury N R 2017 Gas-phase broadband spectroscopy using active sources: progress, status, and applications [Invited] *J. Opt. Soc. Am.* B **34** 104–29
[56] Yang X, Zhao X, Yang K, Liu Y, Liu Y, Fu W and Luo Y 2016 Biomedical applications of terahertz spectroscopy and imaging *Trends Biotechnol.* **34** 810–24
[57] Hafez H A, Chai X, Ibrahim A, Mondal S, Férachou D, Ropagnol X and Ozaki T 2016 Intense terahertz radiation and their applications *J. Opt.* **18** 093004
[58] Redo-Sanchez A, Laman N, Schulkin B and Tongue T 2013 Review of terahertz technology readiness assessment and applications *J. Infrared Millim. Terahertz Waves* **34** 500–18
[59] Le Gros M A, McDermott G, Cinquin B P, Smith E A, Do M, Chao W L, Naulleau P P and Larabell C A 2014 Biological soft x-ray tomography on beamline 2.1 at the advanced light source *J. Synchrotron Radiat.* **21** 1370–7
[60] Fuchs S *et al* 2016 Nanometer resolution optical coherence tomography using broad bandwidth XUV and soft x-ray radiation *Sci. Rep.* **6** 20658
[61] Miao J, Sandberg R L and Song C 2012 Coherent x-ray diffraction imaging *IEEE J. Sel. Top. Quantum Electron.* **18** 399–410
[62] Pertot Y *et al* 2017 Time-resolved x-ray absorption spectroscopy with a water window high-harmonic source *Science* **355** 264
[63] Tisch J W G 2008 Ultrafast goes ultralong *Nat. Phys.* **4** 350–1

[64] Shiner A D, Trallero-Herrero C, Kajumba N, Bandulet H C, Comtois D, Légaré F, Giguère M, Kieffer J C, Corkum P B and Villeneuve D M 2009 Wavelength scaling of high harmonic generation efficiency *Phys. Rev. Lett.* **103** 073902
[65] Haas J and Mizaikoff B 2016 Advances in mid-infrared spectroscopy for chemical analysis *Annu. Rev. Anal. Chem.* **9** 45–68
[66] Henderson B *et al* 2018 Laser spectroscopy for breath analysis: towards clinical implementation *Appl. Phys.* B **124** 161
[67] Griffiths P R and de Haseth J A 2007 *Fourier Transform Infrared Spectrometry* 2nd edn (New York: Wiley) https://doi.org/10.1002/047010631X
[68] Baker M J *et al* 2014 Using Fourier transform IR spectroscopy to analyze biological materials *Nat. Protoc.* **9** 1771
[69] Doherty J, Cinque G and Gardner P 2017 Single-cell analysis using Fourier transform infrared microspectroscopy *Appl. Spectrosc. Rev.* **52** 560–87
[70] Petrov V 2015 Frequency down-conversion of solid-state laser sources to the mid-infrared spectral range using non-oxide nonlinear crystals *Prog. Quantum Electron.* **42** 1–106
[71] Boyd R W 2008 *Nonlinear Optics* 3rd edn (Amsterdam: Elsevier)
[72] Dhillon S S *et al* 2017 The 2017 terahertz science and technology roadmap *J. Phys. D: Appl. Phys.* **50** 043001
[73] Clerici M *et al* 2013 Wavelength scaling of terahertz generation by gas ionization *Phys. Rev. Lett.* **110** 253901
[74] Gaida C, Kienel M, Müller M, Klenke A, Gebhardt M, Stutzki F, Jauregui C, Limpert J and Tünnermann A 2015 Coherent combination of two Tm-doped fiber amplifiers *Opt. Lett.* **40** 2301

Chapter 3

Transition metal doped zinc selenide infrared lasers for ultrafast and intense field science

Yi Wu[1], Esben Witting Larsen[1], Fangjie Zhou, Lifeng Wang, Alphonse Marra, Daniel Schade, Jialin Li and Zenghu Chang

Isolated attosecond x-ray pulses in the water window (282–533 eV) are powerful tools for observing charge dynamics in organic molecules and condensed matter materials containing carbon, nitrogen, and oxygen. The development of such light sources based on high harmonic generation requires mJ-level, few-cycle, carrier-envelope phase stable driving lasers centered at the short-wave infrared or mid-wave infrared spectral region. Recent breakthroughs in Cr^{2+}:ZnSe chirped-pulse amplifiers centered at 2.5 μm have paved the way to bring tabletop attosecond x-ray sources to university laboratories. Fe^{2+}:ZnSe lasers centered at 4 μm may further extend the x-ray photon energy range beyond 1 keV. In this chapter unique schemes for producing microjoule, broadband seed pulses based on intra-pulse difference generation and optical parametric amplification are introduced. The design and implementation of a multi-pass amplification scheme to achieve 4 mJ, 44 fs pulses at 2.5 μm with polycrystalline Cr^{2+}:ZnSe is described.

3.1 Introduction

The advent of chirped-pulse amplified lasers in the 1980s enabled three decades of transformative advances in intense field science [1]. As pointed out by a 2018 National Academies of Sciences report and a recent Bright Light Initiative report [2, 3], a second laser revolution is currently underway with pulsed high-power lasers. Femtosecond lasers with 10 PW or higher are being developed [4, 5]. The center wavelengths of these lasers are at 0.8–1 μm, and utilize Ti, Yb, and Nd doped materials as the gain media. These near-infrared (NIR) wavelengths are neither

[1] Contributed equally

doi:10.1088/978-0-7503-2536-3ch3

magical nor optimal for many intense field applications. They are chosen because there are mature laser materials at these wavelengths.

Moving from the NIR to the short-wavelength infrared (SWIR, 1.1–3 μm) and the mid-wave infrared (MWIR, 3–8 μm) provides the opportunity to take advantage of a list of wavelength scaling laws in intense laser–matter interaction. A very important quantity that determines the outcome of intense laser–matter interaction is the ponderomotive energy, U_p, which is the cycle-averaged kinetic energy a free electron gains in a laser field [6]:

$$U_p\,[\text{eV}] = 9.3I[10^{14}\ \text{W cm}^{-2}]\lambda^2[\mu\text{m}]. \tag{3.1}$$

Here, I and λ are the intensity and wavelength, respectively. Large U_p is required for many intense field applications, which in principle can be achieved by either increasing the laser intensity or the wavelength. However, there are upper limits of intensity for certain nonlinear processes. In those cases, the quadratic wavelength scaling becomes critical. An important example is high harmonic generation (HHG), which was discovered in the late 1980s [7, 8].

HHG is the foundation for tabletop attosecond light sources. The mechanism of HHG is explained by a three-step model [9, 10]. When the intensity of a laser pulse interacting with noble gas atoms reaches $\sim 10^{14}$ W cm^{-2}, an electron can be freed from the atom through tunneling ionization. In the following laser cycle, the electron will be accelerated by the laser field. Finally, the electron recombines with the parent atom and emits a high-photon energy attosecond burst. The photon cutoff energy produced by this extreme nonlinear process, $\hbar\omega_X$, is determined by the highest energy of returning electrons, and is given by:

$$\hbar\omega_X = I_p + 3.17U_p, \tag{3.2}$$

where I_p is the ionization potential of the atom. Attosecond sources based on 1 μm lasers have become so mature that commercial products are now available for user facilities such as the ELI-ALPS in Hungary. In HHG, the laser intensity cannot be too high, typically of the order of 10^{14} W cm^{-2}, to satisfy phase-matching conditions [11, 12].

The index of refraction of noble gas in the SWIR and MWIR regions can be expressed by the Lorentz classical electron oscillator model with a single resonance frequency ω_r. The so-called phase-matching cutoff for a single-cycle driving laser can be estimated by [13]:

$$\hbar\omega_c = I_p + 0.5I_p^{3.5}\lambda_{\text{IR}}^2\left[\ln\left(\frac{2.87I_p3^{2n^*-1}G_{lm}C_{n^*L^*}^2}{-\lambda_{\text{IR}}\ln\left(1-[1.88\omega_r^{-1}\lambda_{\text{IR}}^{-1}]\right)^2}\right)\right]^{-2}. \tag{3.3}$$

The unit of $\hbar\omega_c$ and ionization potential I_p is eV. The unit of laser wavelength λ_{IR} is μm. The values of n^*, G_{lm}, and $C_{n^*L^*}^2$ can be found in [6].

Attosecond sources are powerful tools for studying electron dynamics in molecules and condensed matter materials [14]. Attosecond core-level absorption spectroscopy makes it possible to follow the charge motion from atom to atom in real time. The photon energy range of the 1 μm NIR-driven high flux attosecond source calculated using equation (3.3) is <300eV. For organic molecules, the so-called x-ray water window covering the 280–530 eV range is needed to distinguish carbon from oxygen atoms by their intrinsic absorption edges [6]. For other important molecules and materials, keV attosecond x-rays are required but do not currently exist. Extending the HHG cutoff in the extreme ultraviolet (XUV) region by increasing the driving laser wavelength was first demonstrated with a Ti:Sapphire pumped optical parametric amplifier (OPA) [15]. Empowered by advances in driving lasers with carrier wavelengths between 1.6 and 2.2 μm, significant progress has recently been made in developing attosecond light in the water window [16], but the flux at the oxygen K-edge (530 eV) is still too low for attosecond transient absorption spectroscopy experiments [17]. The scaling law and numerical simulations indicate that phase-matched HHG driven by a 2.5 μm laser can cover the whole water window. Furthermore, 4 μm lasers can generate up to 2.5 keV attosecond x-rays, as shown in figure 3.1.

It is known that the single-atom HHG yield decreases with driver wavelength due to quantum diffusion of the electron wave packet [18, 19]. Obtaining sufficient x-ray flux in the water window and beyond for time-resolved experiments is thus a major challenge [20]. An attosecond pulse burst is generated within one optical cycle of the driving laser. The optical period of a 4 μm laser is 13 fs, which makes it feasible to use a few-femtosecond NIR laser to free an electron from the atom and initiate the attosecond generation process. The electron is then accelerated by the MWIR field. The separation of the two steps is a powerful way to increase the photon flux of the x-ray source [21, 22].

By spectrally broadening pulses from an OPA pumped by 0.8–1 μm lasers using a gas-filled hollow-core fiber [23] or by broadband phase matching in an optical parametric chirped-pulse amplifier (OPCPA), two-cycle, mJ-level pulses centered at 1.7–2 μm with 1 kHz or higher repetition rates are now generated in many

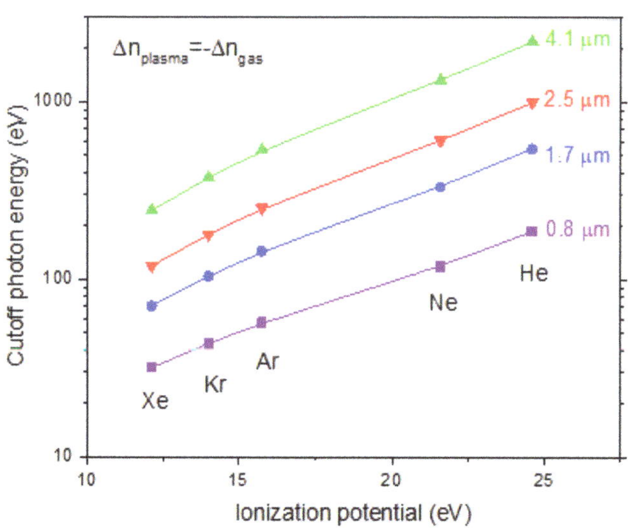

Figure 3.1. Calculated cutoff photon energy of HHG from various inert gases driven by two-cycle lasers at different wavelengths.

laboratories [16]. In principle, this approach can be implemented to achieve high-power ultrafast MWIR pulses [24]. However, this scheme is complex and inefficient.

Transition metal doped chalcogenides have long been promising gain materials for mid-infrared lasing as they provide broad tunable gain media for mid-infrared lasing [25]. In particular, Cr^{2+} and Fe^{2+}-doped chalcogenide lasers have emerged as attractive sources of tunable laser radiation in the 2–5 μm region. Producing large aperture optical quality doped chalcogenides crystals has, however, proven very difficult.

Nevertheless, recent progress in a number of areas, such as broadband, low dispersion, dielectric mirrors, high quality fluoride based transmissive optics, and large aperture polycrystalline chalcogenide based crystals have enabled mJ-class, femtosecond infrared lasers. So far only a few groups have demonstrated mJ-level femtosecond Cr^{2+}:ZnSe lasers [26–28]. The first of its kind was developed in 2016 when Slobodchikov and co-workers demonstrated a three-stage 1 kHz, 1 mJ, 184 fs system [26]. This system consisted of a Cr^{2+}:ZnSe oscillator followed by chirped-pulse amplification (CPA) in a regenerative amplifier pumped by a Q-switched holmium-doped yttrium lithium fluoride (Ho:YLF) laser followed by two single-pass amplification stages [29]. This system has subsequently been compactified into a single booster amplification stage with a footprint of 1.7 m × 1.2 m and an output of 1.5 mJ with a pulse duration of 138 fs [30]. Last year Leshchenko and co-workers presented a 7 mJ, 101 fs, 1 kHz system in an OPA seeded single-stage Cr^{2+}:ZnSe amplifier. This system was subsequently post-compressed to 6.2 mJ, 39 fs through nonlinear compression in a series of CaF_2 lenses and a YAG plate [28].

At the University of Central Florida we have also explored the possibility of utilizing Cr^{2+}:ZnSe as a gain medium for high-energy mid-infrared ultrashort lasers [27, 31]. Femtosecond Fe^{2+}:ZnSe CPA centered at 4.4 μm has been demonstrated, which delivers mJ-level pulses at 10 Hz repetition rate [32].

3.2 Material properties

The spectroscopic characteristics of transition metal doped zinc selenide materials have been studied for more than half a decade [33–36]. A free Cr^{2+} ion has a $[Ar]3d_4$ ground state configuration, while a Fe^{2+} ion has a $[Ar]3d_6$ ground state configuration. The ^5D cation quintet states of Cr^{2+} and Fe^{2+} are in the tetrahedral crystal field of a zinc blende split into a ^5E quintet and a $^5T_2$ quintet. For chromium-doped zinc selenide $^5T_2$ is the ground state and ^5E is the first excited state, while the reverse is the case for iron doping. The first few energy levels of Cr^{2+}:ZnSe and Fe^{2+}:ZnSe are shown in table 3.1. The energy gap between the ground state and the first excited state is smaller for iron ions compared with chromium ions, thus the absorption and emission occur at longer wavelengths.

Cr^{2+}:ZnSe and Fe^{2+}:ZnSe were identified early on as potential gain media for lasers operating at 2.5 μm and 4.5 μm, respectively [37]. The first Cr^{2+}:ZnSe and Fe^{2+}:ZnSe lasers were both first demonstrated in the late 1990s at the Lawrence Livermore National Laboratory [38–40]. The properties are summarized in table 3.2 [41], where the properties of Ti:Sapphire are included for comparison.

Table 3.1. Ground state, and first two excited states for iron and chromium-doped zinc selenide, and their excitation energies [35].

Doping	Ground state	First excited state	Second excited state
Cr^{2+}	($^5T_2$)	0.685 eV (^5E)	1.606 eV ($^3T_2$)
Fe^{2+}	(^5E)	0.365 eV ($^5T_2$)	1.26 eV ($^3T_1$)

Table 3.2. Properties of Cr^{2+}:ZnSe, Fe^{2+}:ZnSe, and $Ti^{3+}:Al_2O_3$ [41]. Values are provided at 300 K unless otherwise specified.

Medium	Cr^{2+}:ZnSe	Fe^{2+}:ZnSe	$Ti^{3+}:Al_2O_3$
λ_{ab}	1.77 μm	3.1 μm	0.488 μm
$\Delta\lambda_{ab}$	0.35 μm	1.37 μm	0.12 μm
σ_{ab}	1.1 Mb	0.97 Mb	0.065 Mb
λ_{em}	2.45 μm	4.35 μm	0.795 μm
$\Delta\lambda_{em}$	0.86 μm	1.61 μm	0.3 μm
σ_{em}	1.3 Mb	2.5 Mb	0.45 Mb
τ_f	5.5 μs	80 μs at 77 K	3.2 μs
K	18 Wm^{-1} K^{-1}	18 Wm^{-1} K^{-1}	28 Wm^{-1} K^{-1}
dn/dT	7×10^{-5} K^{-1}	7×10^{-5} K^{-1}	1.2×10^{-5} K^{-1}
C_V	1.65 mJ cm^{-3}	1.65 mJ cm^{-3}	3.1 mJ cm^{-3}

The absorption and stimulated emission cross section values of Cr^{2+}:ZnSe and Fe^{2+}:ZnSe, $>1 \times 10^{-18}$ cm^2 are larger than that of Ti:Sapphire at 800 nm. The full width at half maximum (FWHM) of the emission cross section curve of Cr^{2+}:ZnSe is 0.86 μm, thus in principle supporting 12 fs transform-limited pulses [42]. The upper-level lifetime of Cr^{2+}:ZnSe is 5.5 μs at room temperature. Consequently, amplifiers can be pumped by Q-switched nanosecond lasers. When pumped by mJ-class Ho: YAG lasers at 2.09 μm (0.59 eV) the quantum defect ($q = \lambda_{em}/\lambda_{abs}$) of the fluorescence is as high as ~83%. In comparison $Ti^{3+}:Al_2O_3$ are normally pumped at around 530 nm yielding a quantum defect of ~66%. When considering the high quantum yield in conjecture with both a thermo-optic coefficient (dn/dT) and volumetric heat capacity (C_V) comparable to that of $Ti^{3+}:Al_2O_3$, it is expected that the thermal lensing for both types of amplifiers are roughly equivalent. For Cr^{2+}:ZnSe, slope efficiencies of continuous-wave lasers and femtosecond pulsed amplifiers of 73% and 30%, respectively, have been demonstrated [28, 43] rivaling or superseding the performances of their $Ti^{3+}:Al_2O_3$ equivalent systems [44, 45]. The high efficiency is achieved by the low quantum defect and the limited impact of excited state absorption, which otherwise leads to losses in efficiency [46]. The excited state absorption is low in Cr^{2+}:ZnSe due to the requirement of spin-forbidden transitions and the large detuning (see table 3.1). Finally, the thermal conductivity, K, of zinc selenide is nearly as high as the conductivity of sapphire, which is critical for high repetition rate and high pulse energy operation.

The luminescence lifetime of Fe^{2+}:ZnSe in the 15–220 K temperature range was studied in [40]. A maximum lifetime of 105 μs was found to occur at a cryogenic temperature of 120 K. The lifetime drops to 370 ns at room temperature as a result of multiphonon quenching [47]. The short upper-state lifetime at room temperature means that short pulsed nanosecond pump lasers are required. Such nanosecond pumps with wavelengths near the 3 μm absorption peak of the Fe^{2+}:ZnSe crystal are not commercially available at the present time. At cryogenic temperature, the crystal can be pumped by gain-switched Er:YAG lasers at 2.94 μm. The stimulated emission cross section as a function of wavelength at 77 K peaks at ~4.1 μm. The FWHM is ~0.5 μm, thus in principal supporting ~50 fs pulses (<4 cycles). The quantum defect of the fluorescence is ~71% when pumped by the 2.94 μm lasers.

3.3 Thermal lensing

The energy deposited into a laser rod can lead to a phenomenon known as thermal lensing [48]. In thermal lensing the laser crystal itself acts as a refractive optic due to thermal distortions of the crystal. The largest contribution to thermal lensing arises from the temperature dependence of the refractive index.

For simplicity we assume that the crystal consists of a cylindrical rod with length L and radius a. The crystal is assumed to be radially cooled to room temperature and axially pumped. In this case the focal length f_{therm} of the thermal lensing may be approximated by an analytical expression as follows [49]:

$$f_{\text{therm}} = \frac{1}{(n-1)} \times \frac{a^3 \pi r_p C_v(T_c)}{\left.\frac{\partial n}{\partial T}\right|_{T_C} E_p[1 - \exp(-\gamma L)]\left[1 - E_{\text{extract}}/E_p\right]} \times \left(\sum_{n=1}^{\infty} \frac{J_1(\alpha_n r_p/a)}{J_1^2(\alpha_n)} \frac{\exp[-\kappa \alpha_n^2/(\nu a^2)]}{1 - \exp[-\kappa \alpha_n^2/(\nu a^2)]} \right)^{-1}, \tag{3.4}$$

where n is the refractive index, $\frac{\partial n}{\partial T}|_{T_C}$ is the thermo-optic coefficient, r_p is the pump laser radius, and $C_v(T_C)$ is volumetric heat capacity of the laser crystal at the center temperature T_C. E_p is the pump energy, and E_{extract} is the optically extracted energy. γ is the linear absorption coefficient of the pump. ν is the repetition rate of the pump laser. J_1 is the Bessel function of the first kind, α_n are the roots of the Bessel function $J_0(\alpha)$. Finally, κ is the thermal diffusivity.

Using this method we find the thermally induced focal length of 12 cm for the conditions shown in table 3.3. Our numerical results show that it is feasible to build high-powered Cr:ZnSe lasers in the MWIR using crystals water-cooled to room temperature by active management of the thermal lensing through the deployment of focusing optics.

Table 3.3. Parameters used for calculating thermal lensing for 2.5 μm light at 300 K.

E_p	E_{extract}	r_p	ν	n	κ	γ	a	L
25 mJ	6 mJ	1 kHz	1.5 mm	2.44	1.15×10^{-7} m^2 s^{-1}	5.7 m^{-1}	1.5 cm	5 cm

3.4 Seed pulse generation

Carrier-envelope phase stable, few-cycle, mJ-level lasers centered at 800 nm are commercially available, and have been widely used to generate isolated attosecond pulses [50]. These lasers are seeded with nanojoule-level pulses from carrier-envelope offset frequency-locked Ti:Sapphire oscillators [51]. Although mode-locked femtosecond Cr^{2+}:ZnSe and Fe^{2+}:ZnSe oscillators have been demonstrated [42, 52], locking the CEP of the pulses is still a challenge. Furthermore, the pulse duration at 4.4 μm is rather long, 732 fs [52]. There are two alternative schemes to seed the Cr^{2+}:ZnSe and Fe^{2+}:ZnSe CPA lasers, intra-pulse difference frequency generation (IDFG), and OPA driven by NIR CPA lasers. Through either of these techniques the seed pulse energy can reach microjoule levels [53], and the carrier-envelope phase can be passively stabilized [54].

3.4.1 Intra-pulse difference frequency generations

Few-cycle, mJ-level white-light pulses covering the spectral range from 550 nm to 950 nm can be generated by coupling the output of a femtosecond Ti:Sapphire CPA laser to a noble gas filled hollow-core fiber [51]. By mixing two pulses centered at two different wavelengths in nonlinear crystals such as bismuth borate (BiB_3O_6, commonly abbreviated BIBO) or potassium titanyle arsenate ($KTiOAsO_4$, commonly abbreviated KTA), a long wavelength pulse can be generated by difference frequency generation (DFG). When the two pulses, normally referred to as the pump and signal pulses, originate from the same white-light pulse the generated DFG pulse will be CEP stable [55]. When a single pulse is sufficiently broadband, it is possible to perform DFG between different spectral parts of the pulse without the use of an interferometric setup through a scheme known as IDFG.

In figure 3.2 our setup for IDFG is shown. Millijoule-level, 30 fs pulses supplied by a Ti:Sapphire laser system are compressed to few-cycle pulses by a gas-filled hollow-core fiber and a set of chirped mirrors and are subsequently focused into a BIBO crystal where the IDFG is performed. Due to the high intensity other unwanted nonlinear effects such as filamentation may occur when the pulse is focused. Therefore, in order to increase the efficiency of the IDFG and to reduce the effect of filamentation, a bandpass filter is used to block the 800 nm spectrum, which does not contribute to the IDFG process. The pulse energy is reduced from 700 μJ to 300 μJ and the resultant spectrum is shown in figure 3.3(a). IDFG between the 650 nm and the 840 nm band generates an idler spectrum that spans from 2100 nm to 3200 nm and peaks at 2500 nm (figure 3.3(b)). This part of the spectrum can be used to seed the Cr^{2+}:ZnSe CPA. The IDFG also induces a number of cascaded second-order

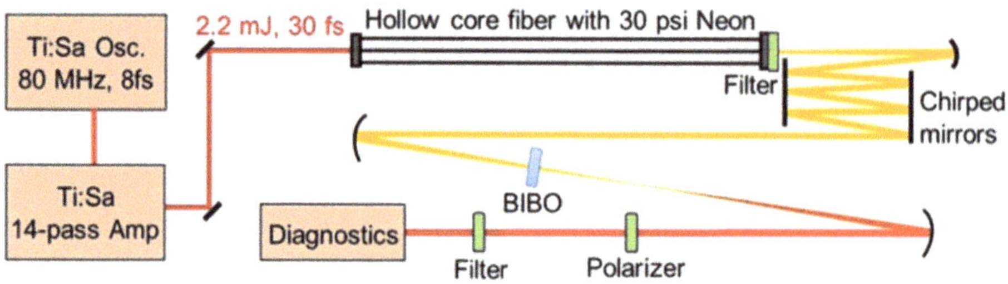

Figure 3.2. Experimental setup for intra-pulse DFG. This figure was reproduced from [53] under the Creative Commons Attribution 4.0 International License: http://creativecommons.org/licenses/by/4.0/. With permission of Springer.

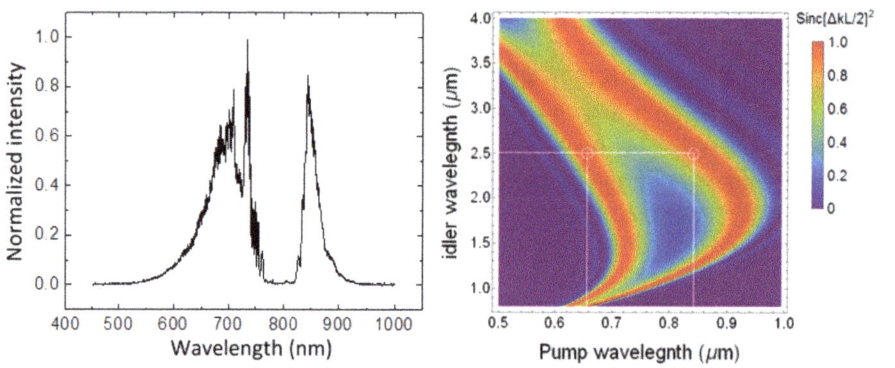

Figure 3.3. Left: The seed laser spectrum used for IDFG. Right: The phase-matching efficiency for different pump and idler wavelengths in BIBO.

nonlinear processes which further extends the spectrum. For example, DFG between the 2400 nm idler and the 840 nm may generate a new pulse at a wavelength of 1300 nm. While the 2400 nm pulse is inherently CEP-phase stable the 840 nm pulse is not, thus any phase-noise in the 840 nm pulse is transferred to the 1300 nm pulse by the DFG process and the relative phase between the two are locked. When the pulses are strong enough yet another DFG process may occur between the 1300 nm pulse and the strong 840 nm resulting in another pulse being generated at 1.7 μm, as shown in figure 3.4(a). As the relative phase between the 840 nm and the 1300 nm pulses is locked the 1.7 μm pulse will inherently become CEP-phase stable, and may be used in seeding a CEP stable, few-cycle OPCPA centered at 1.7 μm [56].

As mentioned above, the output spectrum from the IDFG can be further broadened through filamentation. The filament is formed due to the high peak intensity of the pump pulse. The spectrum of the SWIR to MWIR spanning pulse is thus broadened through the joint effects of self-phase modulation (SPM) and cross-phase modulation (XPM) inside the filament. Figure 3.4 compares the performance with and without the aforementioned bandpass filter and whether a filament is observed in the air after the BIBO. As shown in figure 3.4(a) and (b), the filament leads to a minor extension of the spectrum further into the MWIR (from 3 μm to approximately 3.5 μm) at the cost of distortion of the beam profile (figure 3.4(c)–(e)).

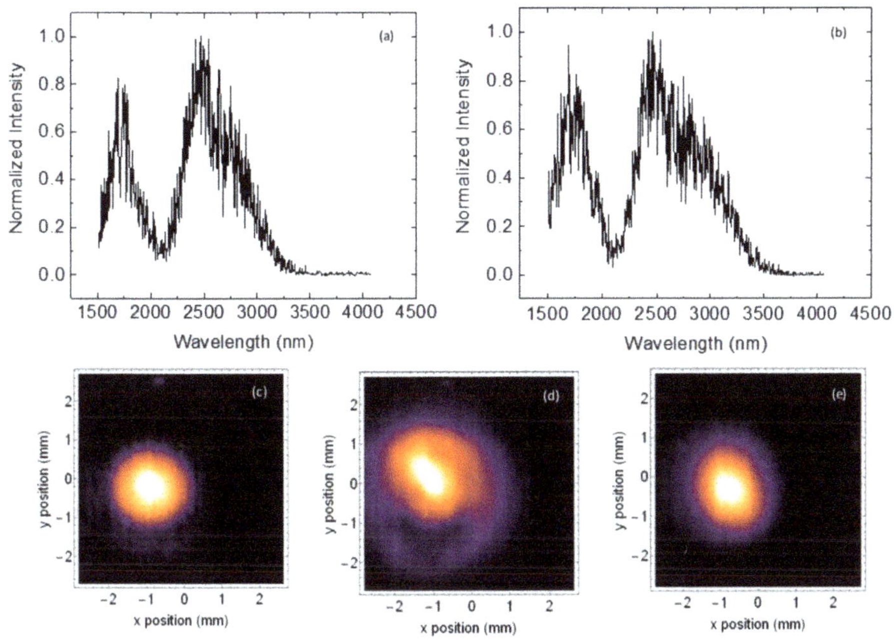

Figure 3.4. The SWIR to MWIR spectrum recorded 100 cm behind the BIBO when (a) there is no filament after the BIBO and (b) a filament is formed in air behind the BIBO. The beam profile 100 cm behind the BIBO when (c) no filament in air is observed. The beam profile when a filament is observed without the bandpass filter (d) and with bandpass filter (e).

Moreover, instability in the output power, profile, and spectrum were observed when the filament was present.

Another trade-off comes from the crystal thickness. A thicker crystal increases the gain but reduces the spectrum bandwidth through gain narrowing, which is determined by the phase-matching bandwidth. Although the transmission of BIBO drops for wavelengths longer than 2.5 μm, it does not significantly affect the bandwidth of the output spectrum since the crystal is thin (~1 mm). Depending on the purpose of the output spectrum, the crystal thickness can vary. High-energy (8 μJ) seed pulses spanning from 1.8 to 3.5 μm have been generated from a combination of the cascaded second-order nonlinear processes in a single 0.8-mm BIBO crystal and the filamentation in air. These pulses are intended to be used to seed a $ZnGeP_2$ OPA capable of delivering high energy few-cycle phase stable idler pulses covering 5 to 10 μm [53, 57]. 12 μJ energy is produced using the same setup with a 1.2 mm BIBO crystal to seed the Cr:ZnSe CPA. Since the spectrum is initiated from the IDFG process, it is expected to have passively stable carrier-envelope phase. The spectrum of the seed pulse is shown in figure 3.4(a). In addition to providing passive CEP stability, seeding the amplifier with pulses from IDFG has two advantages compared with Cr^{2+}:ZnSe and Fe^{2+}:ZnSe mode-locked oscillators. The spectrum is much wider than that from the oscillators. It is also >100 times more intense than the oscillator pulses. As a result, pre-amplification from nJ to μJ is not necessary, which reduces the effects of gain narrowing.

3.4.2 Optical parametric amplifiers

Microjoule-level pulses centered at ~2.5 μm can also be generated from commercially available OPAs and have been deployed in seeding Cr^{2+}:ZnSe CPA lasers [28]. In this work a commercial OPA, Orpheus-HP, from Light-conversion that delivers 2.4 μm, 2 μJ pulses at 75 kHz repetition rate was characterized. The OPA utilizes a turn-key pump laser with a wavelength of 1026 nm, and 80 μJ at a repetition rate of 75 kHz.

The measured spectrum and interferometric autocorrelation traces of the OPA are shown in figure 3.5. The FWHM of the spectrum is about 330 nm, and the corresponding Fourier-transform-limited pulse duration is 27 fs. For pulses with Gaussian temporal shape the pulse duration, τ_p, is related to the width of autocorrelation trace, τ_{ACF}, by:

$$\tau_p = \tau_{\text{ACF}}/\sqrt{2}. \tag{3.5}$$

Although ACF trace is far from Gaussian we can still estimate that the pulses out of the OPA have a duration of around 81 fs. By adding dispersive materials to the beam, the chirped 2.4 μm laser pulses are compressed to a near transform-limited duration of 32 fs.

The near-field and far-field beam profiles were measured with a thermal camera and the results are shown in figure 3.6. The most intense part of near field is close to a Gaussian beam. As observed in figure 3.6(c) the beam can be focused to a near-to-diffraction-limited spot, thus indicating that there is no significant wavefront distortion.

3.5 Dispersion control

In order to prevent damage of the laser media during the amplification process CPA is usually deployed [58]. In CPA the seed pulse is stretched from approximately 50 fs to 300 ps prior to amplification. The chirped pulse is then amplified and subsequently compressed back down to the seed duration thereby enabling the high peak intensities.

When dealing with dispersion of an ultrashort pulse it is advantageous to deal with the electric field in the spectral domain. In the spectral domain an ultrashort pulse can be described as follows

$$\vec{\mathcal{E}}(\vec{r}, \omega) = \hat{\ell}\mathcal{E}_0(\vec{r})A(\omega)e^{i\left(\omega t-\vec{k}\cdot\vec{r}-\phi(\omega)\right)} + \text{c.c.}, \tag{3.6}$$

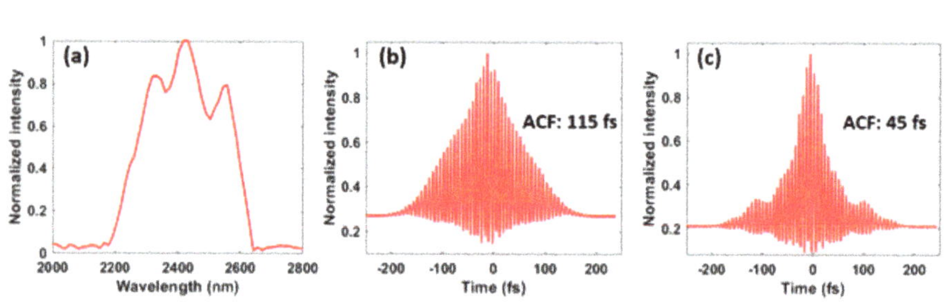

Figure 3.5. Measured spectrum (a) and autocorrelation function traces of a 2.4 μm laser before (b) and after (c) compression. Furthermore, the FWHMs of the ACF are annotated in (b) and (c).

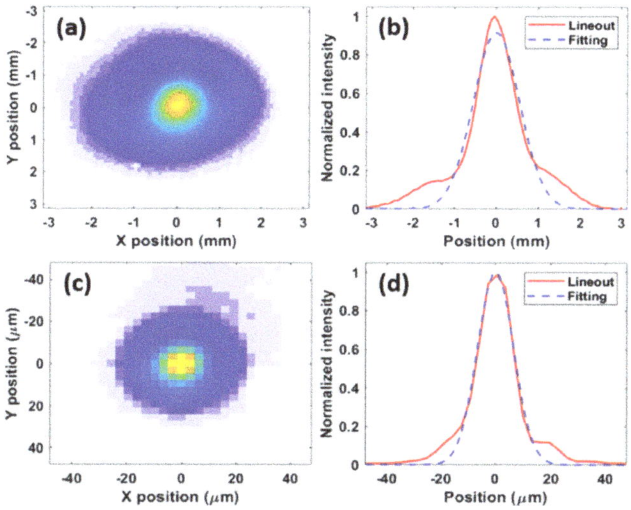

Figure 3.6. (a) Near-field beam profile at 300 mm from the exit port. (c) Beam profile at the focus of a 15 mm focal length lens. (b) lineout of (a) at y = 0 mm. (d) lineout of (c) at y = 0 mm.

here $\hat{\ell}$ denotes the polarization vector, ω the frequency, t the time, $\vec{k}$ the wave vector, $A(\omega)$ and $\mathcal{E}_0(\vec{r})$ are the spectral and spatial envelope functions, respectively, and finally $\phi(\omega)$ denotes a spectral dependent phase.

It is often convenient to approximate $\phi(\omega)$ by a Taylor series around the carrier frequency ω_0 as:

$$\begin{aligned}\phi(\omega) = \underbrace{\phi(\omega_0)}_{\text{CEP}} + \underbrace{\frac{\partial\phi}{\partial\omega}\bigg|_{\omega_0}}_{\text{GD}}(\omega - \omega_0) + \underbrace{\frac{\partial^2\phi}{\partial\omega^2}\bigg|_{\omega_0}}_{\text{GDD}}\frac{(\omega - \omega_0)^2}{2!} \\ + \underbrace{\frac{\partial^3\phi}{\partial\omega^3}\bigg|_{\omega_0}}_{\text{TOD}}\frac{(\omega - \omega_0)^3}{3!} + \cdots.\end{aligned} \tag{3.7}$$

The first term is the absolute phase of the pulse, which is usually referred to as the carrier-envelope phase. The second term is the first-order phase and describes the group delay (GD) of the pulse. The third term is the second-order phase and is called the group delay dispersion (GDD). The fourth and fifth order terms are called the third-order dispersion (TOD) and fourth-order dispersion (FOD), respectively.

In our setup at UCF we utilized a double-pass Öffner triplet type grating stretcher [59, 60] for the stretching of the seed, and double-pass of a pair of parallel gratings as the compressor as shown in figure 3.7.

The dispersion of these optical setups are governed by the optical beam path through the systems. The spectral dependent optical path length, P, through the compressor is given by [61]

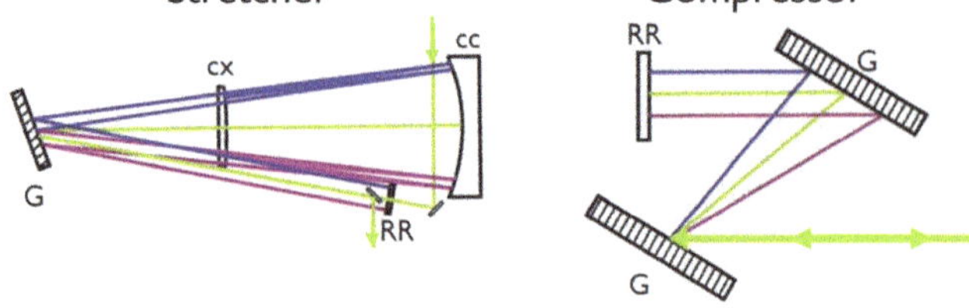

Figure 3.7. Schematic of our stretcher and compressor layouts. RR: Retroreflector, G: Grating, cx: convex mirror, cc: Concave mirror.

$$P(\lambda) = \frac{2L_g}{\cos\theta}(1 + \cos(\gamma - \theta)), \tag{3.8}$$

where λ is the wavelength, L_g is the grating separation, γ is the incidence angle, and θ is the diffraction angle, which may be calculated using the grating diffraction equation

$$\theta = \arcsin\left(\frac{m\lambda}{d} - \sin\gamma\right), \tag{3.9}$$

where d is the grating pitch width and m is the diffraction order. Typically the first-order diffractions, $|m| = 1$, are used due to higher efficiency. The phase shift of the compressor, $\phi_C(\omega)$, may be calculated as [62]:

$$\phi_C(\omega) = P(\omega)k(\omega), \tag{3.10}$$

where $P(\omega)$ may be found from equation (3.8) by utilizing $k = 2\pi/\lambda = \omega n(\omega)/c$. Due to the arbitrary choice of absolute phase the grating phase shift can be written in many different forms, but most compactly as [62, 63]:

$$\phi_C(\omega) = \frac{4\omega L_g}{c}\left[1 - \left(\frac{2\pi c}{\omega d} - \sin\gamma\right)^2\right]^{1/2}. \tag{3.11}$$

Following equation (3.7) we can calculate the dispersion of the compressor up to the FOD by consecutive differentiation of equation (3.11). Martinez *et al* demonstrated that by placing a telescope between a pair of parallel gratings the sign of dispersion could be inverted [62, 64], and apart from aberrations introduced by the telescope the magnitude of dispersion is the same. Thus, a stretcher consisting of the same type of gratings and a telescope may be used to stretch the pulse. In the Öffner triplet stretcher the telescope consists of a pair of concentric spherical mirrors and a single grating in near-Littrow configuration onto which the beam is imaged. The advantage of the Öffner triplet stretcher is that it is nearly aberration free and the utilization of a single grating circumvents practical issues with the gratings not being parallel [60].

In order to achieve the shortest possible compressed pulse after amplification the compressor should compensate for combined dispersion of the stretcher and material dispersion of the transmissive optics inside the laser chain. Table 3.4 presents the

Table 3.4. Dispersion calculations for material dispersion of 20 mm of CaF_2 and 160 mm ZnSe; a double-passed Öffner stretcher at an incidence angle of 25 degrees and a mirror to grating separation of 20 cm; a double-pass, two-grating compressor with a grating separation of 21.6 cm and angle of incidence of 28.1 degrees. Finally, the residual non-compensated phase of the system is shown.

Optical element	GDD (fs^2)	TOD (fs^3)	FOD (fs^4)
Stretcher	1.18×10^6	-5.99×10^6	5.31×10^7
Material	3.25×10^4	6.67×10^4	9.2×10^3
Compressor	-1.21×10^6	5.93×10^6	-6.02×10^7
Residual phase	$<10^2$	$<10^2$	-8.57×10^6

dispersion calculations of our laser chain up to the fourth order. The material dispersion is calculated using the Sellmeier equations for CaF_2 and ZnSe [65, 66].

The optimum grating separation and incidence angle for the compressor are numerically searched in order to compensate for the combined dispersion introduced by the stretcher and material dispersion of the optics up to the TOD. We observe that a large amount of FOD dispersion remains under these conditions. The calculated FOD dispersion would stretch the pulse duration of our amplified pulses from a transform-limited duration of 43 fs Fourier to approximately 175 fs [31]. Furthermore, strong absorption in water above 2.5 μm would further limit the achievable pulse duration, as shown in figure 3.8. Nevertheless, by experimentally tweaking the incidence angle and separation of the compressor grating we were able find a compromise between GDD, TOD, and FOD, which allowed us to achieve a compressed pulse duration around 70 fs, when purging the system with dry nitrogen achieving $<5\%$ humidity (detector limited) [31].

Vyhlidka *et al* proposed to use the aberration induced dispersion of the telescopes inside the stretcher to precompensate for even FOD [67]. However, due to the difficulty of determining dispersion caused by residual water vapor, we chose instead to deploy a SWIR acousto-optic programmable dispersive filter, DAZZLER, from FASTLITE [68], to precompensate for the residual chirp. The DAZZLER allowed us to achieve a near transform-limited duration of 44 fs. Furthermore, the DAZZLER enables a compensating feedback for slow-drift of the CEP [69].

3.6 CPA experimental demonstration

We have demonstrated two iterations of a femtosecond IDFG-seeded Cr^{2+}:ZnSe laser operating at ~2.5 μm [27, 31]. The two layouts of the two systems are shown in figure 3.9 and 3.10, respectively. In the following we will focus on the newer iteration shown in figure 3.10. The seed pulses centered at 2.5 μm are generated by IDFG as described in section 3.4.1. The femtosecond seed pulses are stretched to ~300 ps in a double-pass aberration-free Öffner triplet stretcher containing a gold-coated reflective grating with groove density of 300 lines/mm. The radius of curvature of the convex and the concave mirror of the Öffner telescope is 1 m and 0.5 m, respectively. After the stretcher, the seed pulse has a pulse energy around 1 μJ, which is two to

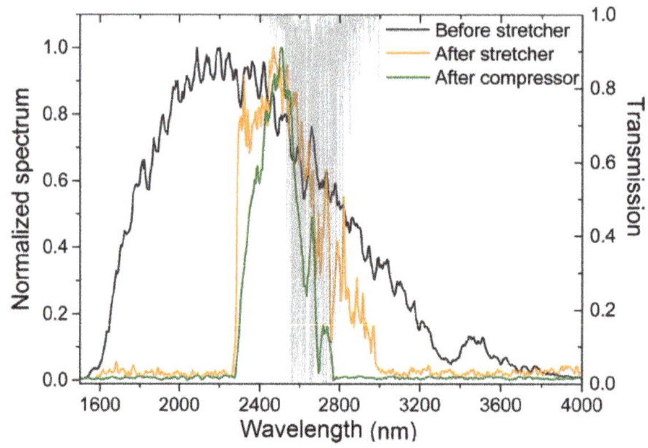

Figure 3.8. Power spectra of our first Cr^{2+}:ZnSe CPA system [27]. The black and orange curves show the seed spectrum after the DFG and the stretcher, respectively. The green curve presents the final output laser spectrum after compressor. Finally, the gray lines show the transmission through 1 cm of pure water vapor. Adapted with permission from [27]. © The Optical Society.

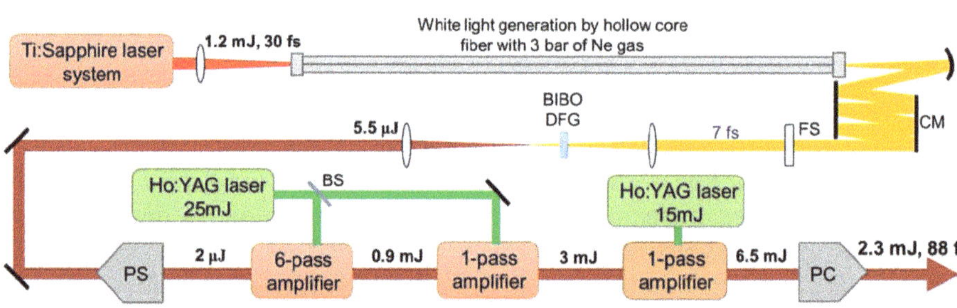

Figure 3.9. The first experimental realization of a high-power a Cr^{2+}:ZnSe laser developed at the University of Central Florida. Adapted with permission from [27]. © The Optical Society.

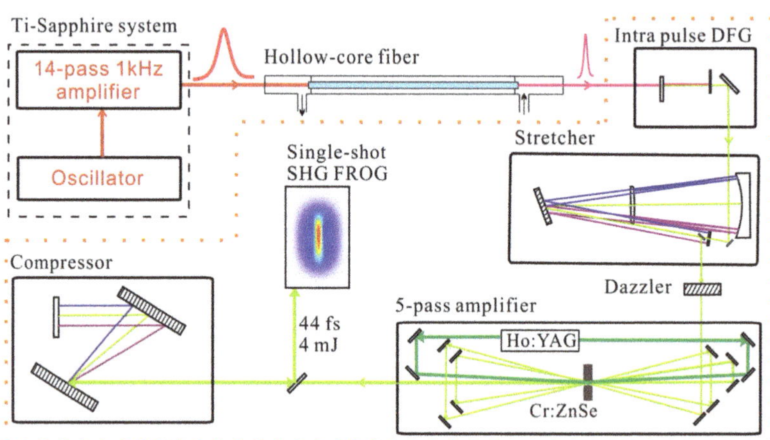

Figure 3.10. The second experimental realization of a high-power Cr^{2+}:ZnSe laser developed at the University of Central Florida. This figure was adapted from [31] under the Creative Commons Attribution 4.0 International License: http://creativecommons.org/licenses/by/4.0/. With permission of Springer.

three orders of magnitude higher than that from a mode-locked oscillator. A SWIR acousto-optic programmable dispersive filter, DAZZLER, is deployed after the stretcher to precompensate the high-order phase accumulated in the amplifier. The DAZZLER further reduces the seed energy to around 0.1 μJ. The 2.5 μm pulses are subsequently directed into a five-pass amplifier. A multi-pass configuration, instead of a regenerative amplification, was chosen to reduce the dispersion of optics.

The gain medium is an anti-reflection coated, 40-mm-long, rectangular polycrystalline Cr^{2+}:ZnSe crystal (IPG Photonics). The crystal height and width are 5 mm and 10 mm, respectively. The crystal is water-cooled to 18°C. 25 mJ, 50 ns, 2.09 μm pulses from a Ho:YAG laser (IPG Photonics) are split in two by a 50/50 beamsplitter and used to pump the crystal from both ends. 75% of the pump energy is absorbed at the full pumping power. The $1/e^2$ diameter of the pump and seed beam is 3.2 mm and 2 mm, respectively. The pulse energy after the fifth pass is 5.9 mJ, thus an absorbed energy to amplified energy ratio of 31% is achieved. A lens was placed in each pass to compensate for the thermal lensing inside the crystal. The seed and output spectra of the system are presented in figure 3.11. The output remains broadband due to the low gain factor ($\sim 10^3$) of the system compared with nanojoule seeded CPAs with gain factors around 10^6. Furthermore, the entire system was built inside a sealed box, and was purged with dry nitrogen during operation to eliminate the strong absorption of water vapor beyond 2.5 μm.

The amplified pulses are compressed in a double-pass grating compressor. The groove density of the grating is same as the stretcher. The throughput of the compressor is 68% resulting in an output pulse energy of 4 mJ. The spectral phase of the output pulses was characterized by a home-built single-shot second harmonic frequency resolved optical gating (FROG), which allowed us to optimize the phase error correction by the DAZZLER. The minimized pulse duration was 44 fs as shown in figure 3.12, which is the shortest pulse duration achieved directly from mJ CPA lasers centered at 2.5 μm.

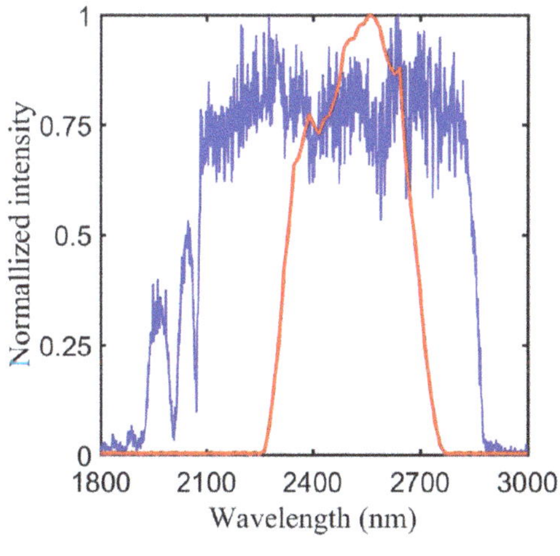

Figure 3.11. Normalized seed (blue) and amplifier (red) spectra of the laser system. This figure was adapted from [31] under the Creative Commons Attribution 4.0 International License: http://creativecommons.org/licenses/by/4.0/. With permission of Springer.

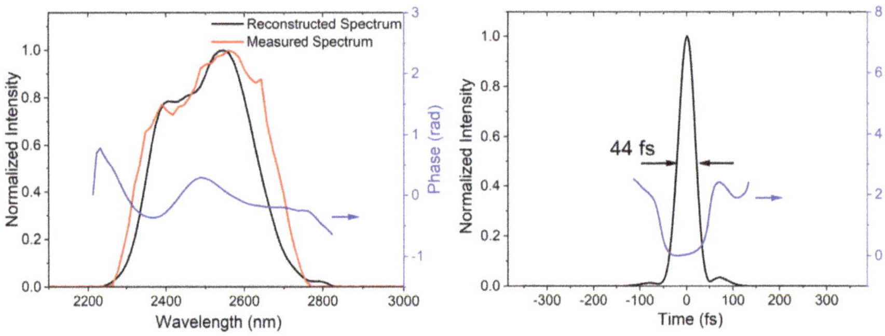

Figure 3.12. Left: Measured (red) and FROG reconstructed (black) spectrum of the output of the Cr^{2+}:ZnSe CPA system as well as spectral phase from the single-shot FROG (blue). Right: Temporal structure of the Cr^{2+}:ZnSe CPA pulses from the FROG reconstruction. This figure was adapted from [31] under the Creative Commons Attribution 4.0 International License: http://creativecommons.org/licenses/by/4.0/. With permission of Springer.

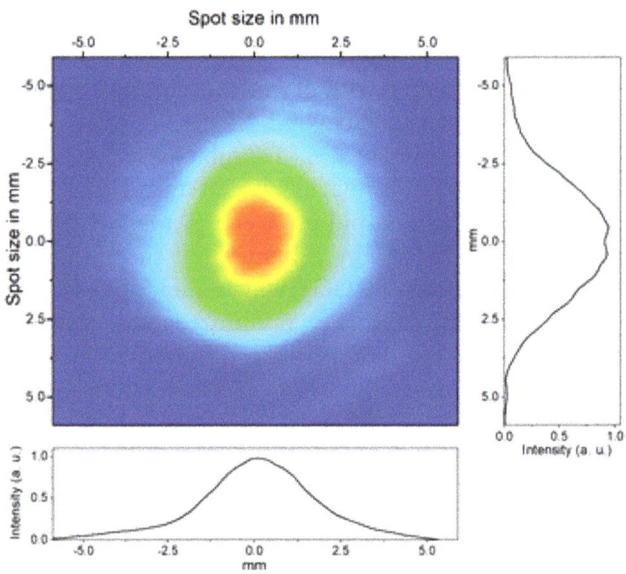

Figure 3.13. Beam profile and lineouts with Cr^{2+}:ZnSe single-stage five-pass CPA system.

The measured near-field and far-field beam profiles are shown in figure 3.13, and are excellent especially considering cryogenic cooling of the gain crystal was not applied, which is often required for Ti:Sapphire lasers with similar output power, pulse energy, and duration.

3.7 Gain calculations

In order to estimate the optimum performance of the amplifier we deploy the methodology described in [70] and outlined here. The pass dependent output fluence, $J_{\mathrm{out}}^{(\mathrm{p})}$, may be calculated as:

$$J_{\text{out}}^{(\text{p})} = J_{\text{sat}} \ln\left\{1 + G_0^{(\text{p})}(\lambda)\left[\exp\left(\frac{J_{\text{in}}^{(\text{p})}}{J_{\text{sat}}}\right) - 1\right]\right\}, \tag{3.12}$$

where $J_{\text{in}}^{(\text{p})}$ is the pass dependent input fluence and J_{sat} is the saturation fluence. $G_0(\lambda)$ is the small signal gain given by:

$$G_0^{(\text{p})}(\lambda) = \exp\left[\sigma_{\text{norm}}(\lambda)J_{\text{stor}}^{(\text{p})}/J_{\text{sat}}\right], \tag{3.13}$$

for which σ_{norm} is the spectrally normalized gain cross section. Additionally between each pass the input fluence is modified by the round trip losses R in the following way:

$$J_{\text{in}}^{(\text{p}+1)} = (1 - R)J_{\text{out}}^{(\text{p})}, \tag{3.14}$$

while the stored fluence is modified as follows:

$$J_{\text{stor}}^{(\text{p}+1)} = J_{\text{stor}}^{(\text{p})} - \left[J_{\text{out}}^{(\text{p})} - J_{\text{in}}^{(\text{p})}\right]. \tag{3.15}$$

It is assumed that the first pass stored fluence is given by

$$J_{\text{stor}}^{(1)} = \eta\alpha J_{\text{pump}}, \tag{3.16}$$

where J_{pump} is the pump fluence, α is the absorption fraction of the pump, and η is the quantum efficiency of the lasing action.

For simplicity we assume that both the pump and the seed beams have uniform transverse beam profiles, i.e., that they are circular tophats with beam radii w_{seed} and w_{pump}, respectively.

The parameters used in our simulations are shown in table 3.5. The results of the spectral reshaping due to gain narrowing are presented in figure 3.14, while the pulse energy of the calculated and experimental pass-by-pass energies are given in table 3.6. We note that this simplified one-dimensional model reproduces overall the performance of the system, but small differences in experimental and theoretical values are obtained, probably due to the non-collinear experimental implementation.

3.8 Post-compression and filamentation for the generation of single-cycle mid-infrared pulses

The shortest pulse duration achieved from a mJ-level Cr^{2+}:ZnSe laser amplifier centered at 2.5 μm is 44 fs [31]. For efficient generation of isolated attosecond x-ray pulses in the water window, the driving laser should approach one optical cycle,

Table 3.5. Parameters used for gain simulations.

E_{seed}	E_{pump}	w_{seed}	w_{pump}	λ_{seed}	$\Delta\lambda_{\text{seed}}$	λ_{pump}	α	R
0.1 μJ	25 mJ	1 mm	1.6 mm	2.45 μm	0.4 μm	2.09 μm	75%	15%

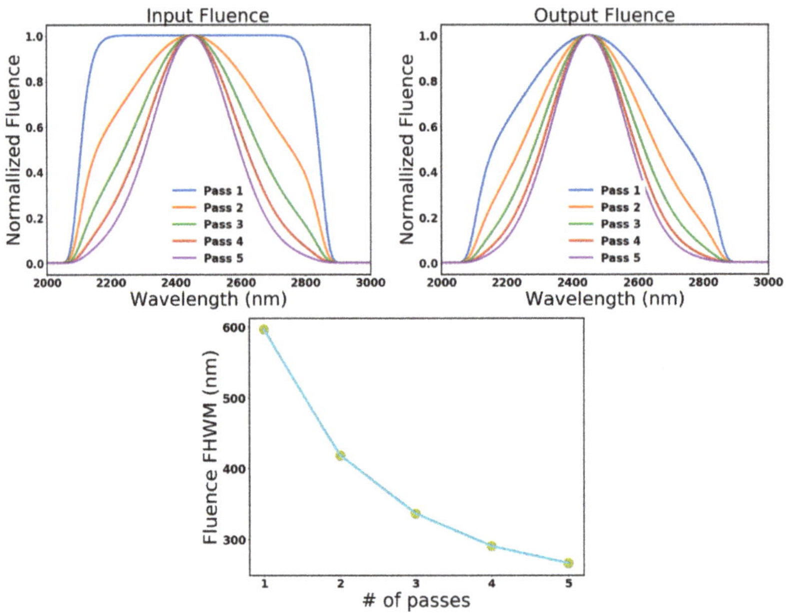

Figure 3.14. Top left: Normalized input spectra for each pass. Top right: Normalized output spectra for each pass. Bottom: FWHM bandwidth of the output the spectral profile.

Table 3.6. Comparison of gain calculation between experimental and numerical results.

Pass	Experimental output energy	Calculated output energy
Seed	<1 μJ	0.1 μJ
1	1.5 μJ	1.2 μJ
2	25 μJ	14.6 μJ
3	350 μJ	180 μJ
4	1.8 mJ	2.3 mJ
5	5.9 mJ	6.6 mJ

which is 8.3 fs for 2.5 μm light; thus requiring a post-compression of the CPA output by a factor 5. In the case of Fe^{2+}:ZnSe, the shortest pulses are 150 fs [32], which correspond to roughly 10 optical cycles at 4.4 μm.

Many approaches have been demonstrated for compressing mJ-level NIR pulses to few-cycles [71], which include spectral broadening in a gas-filled hollow-core fiber, light filamentation, multiple thin solid plates, and multi-pass gas cell, followed by chirped mirrors with negative GDD to compensate for the positive chirp induced by the SPM. Whereas chirped mirrors for SWIR and MWIR are still being developed, the negative dispersion of materials in the long wavelength region can be utilized, although the contrast of the pulses may not be as good as that achieved with chirped mirrors due to the effects of the TOD of the materials. Recently, Leshchenko *et al*

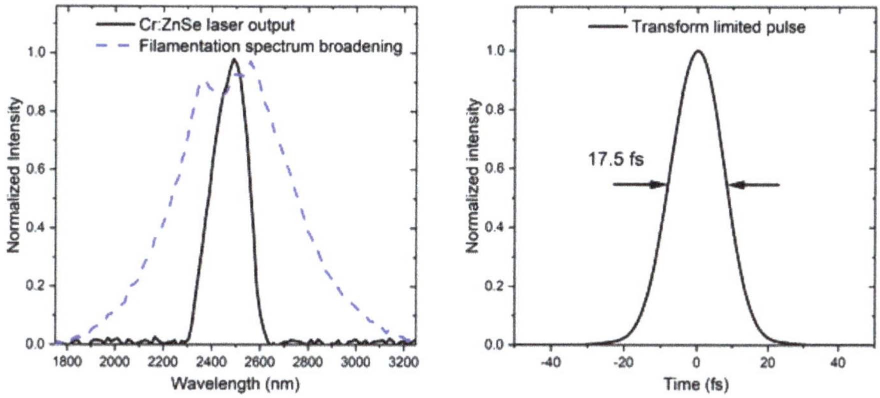

Figure 3.15. Left: Measured spectrum before and after filamentation induced spectral broadening. Right: Calculated transform-limited pulse temporal profile for a filamentation spectrum broadened beam.

compressed the output of a Cr^{2+}:ZnSe CPA centered at 2.4 μm from 101 fs to 39 fs by taking advantage of the SPM and negative GVD of CaF_2 lenses [28].

In our case we have deployed the technique of laser filamentation in air. We observed spectral broadening by focusing the mJ-level pulses in air using a lens with an f-number of 50. The measured spectrum is presented in figure 3.15, which supports 17.5 fs transform-limited pulses. The advantage of this scheme is the potential for high-energy throughput [72].

3.9 Future perspectives

Significant progress has been made in developing femtosecond, mJ-level Cr^{2+}:ZnSe and Fe^{2+}:ZnSe lasers centered at 2.5 and 4.4 μm, respectively. In the next five years, it is expected that the pulse duration of these high pulse energy lasers will be reduced to less than two cycles and their carrier-envelope phase will be stabilized, which is critical for generating single isolated attosecond pulses in the 0.2–2.5 keV x-ray region. The repetition rate may reach 10 kHz or higher, which may require cryogenic cooling of the gain medium. High repetition rate attosecond sources are important for conducting ion-electron coincidence experiments. The peak power at low repetition rate (100 Hz or less) may reach terawatt for relativistic laser–plasma interaction experiments.

It is expected that the emerging high-power MWIR lasers will break many barriers encountered by NIR-driven research. When the intensity is around 10^{12}–10^{13} W cm^{-2}, the index of refraction of gases can be changed by the laser field. Consequently, a light-induced lens can be formed that can focus the laser beam, which is called Kerr self-focusing. The Kerr self-focusing can be balanced by plasma defocusing that results in a light filament in gas. The critical power of Kerr self-focus is proportional to the square of the laser wavelength. This scaling law indicates that it might be possible to transport much higher laser power by a filament in the MWIR than in the NIR region [73, 74]. There is a wide atmosphere window in the 3–5 μm region where the transmission is high, which allows the light filamentation to propagate over a long distance. Therefore, high-power MWIR lasers bear the potential to have a profound impact on electronic countermeasure, remote sensing, directed energy, free-space communication, and other critical defense applications.

When the laser intensity is sufficiently high, gas can be ionized and become plasma. The index of refraction of plasma can be changed by variation of the mass of an electron with speed, which can lead to the so-called relativistic self-focusing. Laser filamentation as a result of the relativistic self-focusing is the first step in accelerating electron pulses to GeV energy. Since the critical power of such process is inversely proportional to the square of the wavelength, less laser power is required for long wavelength lasers. The combination of this and other scaling laws makes it possible to create a large plasma structure for electron acceleration. Such a large bubble generated by MWIR allows one to control the quality of the output electron beam by seeding of electrons from an external source, which is hard to implement with NIR driving lasers.

Attosecond x-ray sources driven by MWIR lasers have the potential to become powerful tools for studying ultrafast magnetism, which may lead to the establishment of future petahertz spintronic devices such as spin transistors and data storage media. It has been demonstrated recently that the magnetic properties of a stack of ferromagnetic layers can be manipulated directly by the oscillations of an intense laser field [75]. The coupled optical and magnetic excitation occurring within one femtosecond was measured by an attosecond time-resolved magnetic circular dichroism at the M-edge of Co (60 eV). The mechanism of the ultrafast demagnetization occurring on sub-optical-cycle time scales is described by optically induced spin and orbital momentum transfer caused by light-field-driven coherent charge relocation. Attosecond x-ray pulses in the 700–900 eV range based on HHG may play an important role in studying electron dynamics in ferromagnetic magnet materials that contain transition metal elements including Fe, Co, and Ni. The absorption edges of these elements are 708.1 eV (Fe L2), 721.1 eV (Fe L1), 778.6 eV (Co L2), 793.8 eV (Co L1), 854.7 eV (Ni L2), and 871.9 eV (Ni L1). Performing magnetic circular dichroism at the L-edges instead of at the lower energy M-edges is often advantageous, since thicker samples can be studied. Furthermore, spin–orbit couplings have a larger energy separation at the L-edges, enabling the decoupling of spin–orbit effects in magneto-absorption studies.

Acknowledgments

This work was supported by the United States Air Force Office of Scientific Research (AFOSR) (FA9550-15-1-0037, FA9550-16-1-0013, FA9550-20-1-0295); Army Research Office (ARO) (W911NF-14-1-0383, W911NF-19-1-0224); Defense Advanced Research Projects Agency (DARPA) (D18AC00011); National Science Foundation (1806575); Defense Threat Reduction Agency (HDTRA11910026). The Defence Science and Technology Laboratory and The Engineering and Physical Sciences Research Council (DSTL/EPSRC: MURI EP/N018680/1). We thank Jon Marangos for his support, and Timur Avni and Bruce Weaver for technical discussions.

References

[1] Mourou G A, Tajima T and Bulanov S V 2006 *Rev. Mod. Phys.* **78** 309–71

[2] 2018 Board on Physics and Astronomy, Division on Engineering and Physical Sciences, The National Academies of Sciences, Engineering and Medicine *Opportunities in Intense Ultrafast Lasers: Reaching for the Brightest Light* (The National Academies Press) https://www.nap.edu/catalog/24939/opportunities-in-intense-ultrafast-lasers-reaching-for-the-brightest-light

[3] Falcone R, Albert F, Beg F, Glenzer S, Ditmire T, Spinka T and Zuegel J 2020 arXiv:2002.09712v2

[4] Gales S *et al* 2018 *Rep. Prog. Phys.* **81** 094301

[5] Lureau F *et al* 2020 10 petawatt lasers for extreme light applications *Solid State Lasers XXIX: Technology and Devices* vol 11259 ed W A Clarkson and R K Shori (San Francisco, CA: International Society for Optics and Photonics (SPIE)) pp 283–93

[6] Chang Z 2011 *Fundamentals of Attosecond Optics* (Boca Raton: CRC Press)

[7] McPherson A, Gibson G, Jara H, Johann U, Luk T S, McIntyre I A, Boyer K and Rhodes C K 1987 *J. Opt. Soc. Am.* B **4** 595

[8] Ferray M, L'Huillier A, Li X F, Mainfray G and Manus C 1988 *J. Phys.* B **21** L31

[9] Schafer K J, Yang B, DiMauro L F and Kulander K C 1993 *Phys. Rev. Lett.* **70** 1599

[10] Corkum P B 1993 *Phys. Rev. Lett.* **71** 1994

[11] Rundquist A, III C D, Chang Z, Herne C, Backus S, Murnane M and Kapteyn H C 1998 *Science* **280** 1412

[12] Popmintchev T, Chen M C, Bahabad A, Gerrity M, Sidorenko P, Cohen O, Christov I P, Murnane M M and Kapteyn H C 2009 *Proc. Natl. Acad. Sci.* **106** 10516–21

[13] Han S, Li J, Zhu Z, Chew A, Larsen E W, Wu Y, Pang S S and Chang Z 2020 Advances in atomic, molecular, and optical physics *Chapter One—Tabletop attosecond X-rays in the Water Window* vol 69 ed L F Dimauro, H Perrin and S F Yelin (New York: Academic) pp 1–65

[14] Leone S R *et al* 2014 *Nat. Photon.* **8** 162–6

[15] Shan B and Chang Z 2001 *Phys. Rev.* A **65** 011804

[16] Ren X 2018 *J. Opt.* **20** 023001

[17] Saito N, Sannohe H, Ishii N, Kanai T, Kosugi N, Wu Y, Chew A, Han S, Chang Z and Itatani J 2019 *Optica* **6** 1542–6

[18] Tate J, Auguste T, Muller H G, Salieres P, Agostini P and DiMauro L F 2007 *Phys. Rev. Lett.* **98** 013901

[19] Shiner A D, Trallero-Herrero C, Kajumba N, Bandulet H C, Comtois D, Legare F, Giguere M, Kieffer J C, Corkum P B and Villeneuve D M 2009 *Phys. Rev. Lett.* **103** 073902

[20] Cousin S L, Di Palo N, Buades B, Teichmann S M, Reduzzi M, Devetta M, Kheifets A, Sansone G and Biegert J 2017 *Phys. Rev.* X **7** 041030

[21] Schafer K J, Gaarde M B, Heinrich A, Biegert J and Keller U 2004 *Phys. Rev. Lett.* **92** 23003

[22] Chang Z 2019 *OSA Continuum* **2** 2131–6

[23] Schmidt B E *et al* 2010 *Appl. Phys. Lett.* **96** 121109

[24] Thiré N, Maksimenka R, Kiss B, Ferchaud C, Bizouard P, Cormier E, Osvay K and Forget N 2017 *Opt. Express* **25** 1505–14

[25] Mirov S B, Moskalev I S, Vasilyev S, Smolski V, Fedorov V V, Martyshkin D, Peppers J, Mirov M, Dergachev A and Gapontsev V 2018 *IEEE J. Select. Top. Quant. Electron.* **24** 1–29

[26] Slobodchikov E, Chieffo L R and Wall K F 2016 High peak power ultrafast Cr:ZnSe oscillator and power amplifier *Solid State Lasers XXV: Technology and Devices* vol 9726 ed W A Clarkson and R K Shori (San Francisco, CA: International Society for Optics and Photonics (SPIE)) pp 17–23
[27] Ren X, Mach L H, Yin Y, Wang Y and Chang Z 2018 *Opt. Lett.* **43** 3381–4
[28] Leshchenko V E, Talbert B K, Lai Y H, Li S, Tang Y, Hageman S J, Smith G, Agostini P, DiMauro L F and Blaga C I 2020 *Optica* **7** 981–8
[29] Slobodchikov E and Moulton P F 2011 1-GW-peak-power, Cr:ZnSe laser *CLEO:2011—Laser Applications to Photonic Applications* (Baltimore, MA: Optical Society of America) p PDPA10
[30] Vasilyev S, Peppers J, Moskalev I, Smolski V, Mirov M, Slobodchikov E, Dergachev A, Mirov S and Gapontsev V 2019 1.5-mJ Cr:ZnSe chirped pulse amplifier seeded by a Kerr-lens mode-locked Cr:ZnS oscillator *Laser Congress 2019 (ASSL, LAC, LS&C)* (Optical Society of America) p ATu4A.4
[31] Wu Y, Zhou F, Larsen E W, Zhuang F, Yin Y and Chang Z 2020 *Sci. Rep.* **10** 7775
[32] Migal E, Pushkin A, Bravy B, Gordienko V, Minaev N, Sirotkin A and Potemkin F 2019 *Opt. Lett.* **44** 2550–3
[33] Vallin J, Slack G, Roberts S and Hughes A 1969 *Solid State Commun.* **7** 1211–4
[34] Baranowski J M, Allen J W and Pearson G L 1967 *Phys. Rev.* **160** 627–32
[35] Fazzio A, Caldas M J and Zunger A 1984 *Phys. Rev.* B **30** 3430–55
[36] Colignon D, Kartheuser E, Rodríguez S and Villeret M 1996 *J. Cryst. Growth* **159** 875–8
[37] Mirov S B, Fedorov V V, Martyshkin D, Moskalev I S, Mirov M and Vasilyev S 2015 *IEEE J. Select. Top. Quant. Electron.* **21** 292–310
[38] DeLoach L D, Page R H, Wilke G D, Payne S A and Krupke W F 1996 *IEEE J. Quant. Electron.* **32** 885–95
[39] Page R H, Skidmore J A, Schaffers K I, Beach R J, Payne S A and Krupke W F 1997 Demonstrations of diode—pumped and grating—tuned znse:cr2+ lasers *Advanced Solid State Lasers* (Orlando, FL: Optical Society of America) p LS6
[40] Adams J J, Bibeau C, Page R H, Krol D M, Furu L H and Payne S A 1999 *Opt. Lett.* **24** 1720–2
[41] Xue Y, Xu X, Su L and Xu J 2020 *J. Synth. Cryst.* **49** 1347
[42] Sorokina I T and Sorokin E 2015 *IEEE J. Select. Top. Quant. Electron.* **21** 273–91
[43] Podlipensky A V, Shcherbitsky V G, Kuleshov N V, Levchenko V I, Yakimovich V N, Mond M, Heumann E, Huber G, Kretschmann H and Kück S 2001 *Appl. Phys.* B **72** 253–5
[44] Kanetake T, Hayashi K, Kadoya H, Inayoshi S, Kataoka S, Sugiki F, Nakajima N, Kobayashi R and Kawato S 2018 High efficiency continuous-wave Ti:Sapphire laser *Conference on Lasers and Electro-Optics* (San Jose, CA: Optical Society of America) p JTu2A.92
[45] Zhang W, Teng H, Wang Z, Shen Z and Wei Z 2013 *Appl. Opt.* **52** 1517–22
[46] Kück S 2013 *Recent Development in Laser Crystals with 3D Ions* (Berlin Heidelberg: Springer) pp 1–28
[47] Fedorov V V *et al* 2006 *IEEE J. Quant. Electron.* **42** 907–17
[48] Koechner W 1970 *Appl. Opt.* **9** 2548–53
[49] Lausten R and Balling P 2003 *J. Opt. Soc. Am.* B **20** 1479–85
[50] Chini M, Zhao K and Chang Z 2014 *Nat. Photon.* **8** 178

[51] Mashiko H, Nakamura C M, Li C, Moon E, Wang H, Tackett J and Chang Z 2007 *Appl. Phys. Lett.* **90** 161114
[52] Pushkin A V, Migal E A, Tokita S, Korostelin Y V and Potemkin F V 2020 *Opt. Lett.* **45** 738–41
[53] Yin Y, Ren X, Chew A, Li J, Wang Y, Zhuang F, Wu Y and Chang Z 2017 *Sci. Rep.* **7** 11097
[54] Komm P, Sheintop U, Noach S and Marcus G 2019 *Opt. Express* **27** 18522–32
[55] Baltuška A, Fuji T and Kobayashi T 2002 *Phys. Rev. Lett.* **88** 133901
[56] Yin Y, Li J, Ren X, Zhao K, Wu Y, Cunningham E and Chang Z 2016 *Opt. Lett.* **41** 1142–5
[57] Yin Y, Chew A, Ren X, Li J, Wang Y, Wu Y and Chang Z 2017 *Sci. Rep.* **7** 45794
[58] Strickland D and Mourou G 1985 *Opt. Commun.* **56** 219
[59] Öffner A 1971 *U.S. Patent* 3,748,015 https://www.lens.org/lens/patent/US_3748015_A
[60] Cheriaux G, Rousseau P, Salin F, Chambaret J P, Walker B and Dimauro L F 1996 *Opt. Lett.* **21** 414–6
[61] Treacy E 1969 *IEEE J. Quant. Electron.* **5** 454–8
[62] Martinez O E, Gordon J P and Fork R L 1984 *J. Opt. Soc. Am.* B **1** 1003
[63] Backus S, Durfee C G, Murnane M M and Kapteyn H C 1998 *Rev. Sci. Instrum.* **69** 1207–23
[64] Martinez O 1987 *IEEE J. Quant. Electron.* **23** 1385–7
[65] Tatian B 1984 *Appl. Opt.* **23** 4477–85
[66] Malitson I H 1963 *Appl. Opt.* **2** 1103–7
[67] Vyhlídka Š, Kramer D, Meadows A and Rus B 2018 *Opt. Commun.* **414** 207–11
[68] Maksimenka R *et al* 2010 *Opt. Lett.* **35** 3565–7
[69] Fordell T, Miranda M, Persson A and L'Huillier A 2009 *Opt. Express* **17** 21091–7
[70] Tokita S, Hashida M, Masuno S, Namba S and Sakabe S 2008 *Opt. Express* **16** 14875–81
[71] Nagy T, Simon P and Veisz L 2021 *Adv. Phys.:* X **6** 1845795
[72] Schmidt C, Pertot Y, Balciunas T, Zinchenko K, Matthews M, Wörner H J and Wolf J P 2018 *Opt. Express* **26** 11834–42
[73] Mitrofanov A V *et al* 2015 *Sci. Rep.* **5** 8368
[74] Tochitsky S *et al* 2019 *Nat. Photon.* **13** 41–6
[75] Siegrist F *et al* 2019 *Nature* **571** 240–4

Chapter 4

100 kHz tunable mid-IR ultrafast sources for high intensity applications

Yoann Pertot, Nicolas Thiré, Raman Maksimenka, Olivier Albert and Nicolas Forget

4.1 Introduction

The remarkable development of high-power picosecond lasers based on Yb:YAG technology [1–8] allows the development of mid-infrared (MIR) parametric sources combining short pulses (<100 fs), high peak intensity (>1 GW), high average power (>10 W) and wide tunability (octave-spanning). Because of the intrinsically low conversion efficiency of non-linear processes, parametric sources were essentially limited by the available energy/power of the so-called 'pump' lasers. The advent of kW-class pump lasers delivering pulse energy of several millijoules at very high repetition rate (100 kHz) removed such a barrier and opened the path to a large set of applications, among which are high-flux X-UV secondary sources. It is interesting to note that the development of picosecond Yb:YAG lasers was not driven by the scientific market but rather by the industrial market of welding, micro-machining and connected applications. Because of the pulse duration and fixed wavelength of these lasers, most scientific applications still require additional non-linear devices to shorten the pulse duration, shift the central wavelength or add some physical properties (carrier-envelope phase control and stability) while keeping the excellent beam profile and energy stability. If such devices or techniques, including opcpa systems or spectral broadening, already exists, their adaptation to the new space of parameters (duration, repetition rate, wavelength) and constraints (average power) remains challenging and is an active field of technological research.

One way to shorten pulse duration is to propagate in gas-filled hollow-core fibers to induce spectral broadening via self-phase modulation [9, 10]. Pulses are then compressed by a set of dispersive mirrors or bulk material. The throughput of such fibers typically reaches 50% but few-cycle pulses may be generated with this

doi:10.1088/978-0-7503-2536-3ch4

technique. However, carrier-envelope-phase (CEP) stability has been managed prior to this broadening stage and the spectral tunability/coverage is very limited. Moreover, the footprint and the technical developments of such devices lead to new challenges.

For the latter reasons, optical parametric chirped-pulse amplification (OPCPA) is often preferred to produce tunable short pulses in the near-infrared (NIR) and mid-infrared (MIR) wavelength range [3, 11–13]. OPCPA relies on a non-linear process called three-wave mixing (figure 4.1), allowing a local and instantaneous[1] energy transfer from an intense wave, called a pump wave, to a weaker seed wave of longer wavelength, called a signal wave. Due to energy conservation, a complementary photon is emitted during the process. This wave is called the idler wave or, depending on the context, the wave generated by frequency difference (DFG). OPCPA is a particular case of optical parametric amplification (OPA) in which the amplified wave is a chirped pulse (CPA, chirped-pulse amplification). It is therefore implicitly assumed in OPCPA that (i) the fields correspond to the domain of optical pulses (0.5–10 μm), (ii) the spectrum of seed/signal pulses is broadband (FTL <100 fs), (iii) the seed/signal pulse is chirped significantly (10–50 time) with respect to its Fourier-transform limit, (iv) the duration of the chirped seed/signal pulses matches that of the pump pulse, (v) the chirp is compensated by a compression stage before the output of the OPCPA.

Although a detailed discussion is out of scope of this chapter, it must be mentioned that the benefits and constraints of CPA in OPAs differ significantly from CPA in laser amplifiers. Let it suffice to state that the three major differences are (i) the absence of gain storage, (ii) the role of chirp and (iii) the requirements on pulse-seed synchronization. These three key differences all derive directly from the instantaneity of the parametric gain. First, in OPCPA, there is no energy storage in the amplification medium: the pump pulse is the only energy reservoir. Second, the seed pulses are stretched not only to lower the peak intensity but also, and above all, to optimize the temporal overlap between the seed and pump pulses. Third, the synchronization between the pump and seed pulses must be accurate and stable to less than a small fraction (<1–10%) of the pump and/or signal pulse durations. In laser amplifiers, these constraints are relaxed by the lifetime of the excited state,

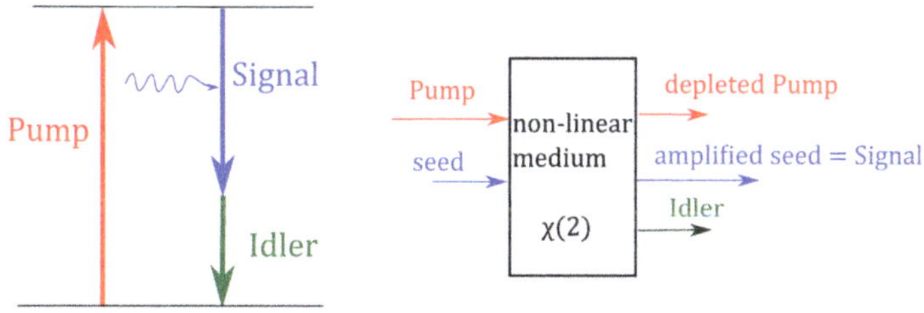

Figure 4.1. Basics of parametric amplification in a non-linear medium based on a $\chi^{(2)}$ process.

[1] Shorter than the optical cycles.

which lies in the μs–ms range for most conventional laser amplifiers (semi-conductor lasers being an exception). If the lack of gain storage mechanism and of associated quantum defect in OPCPA is unquestionably a technological weakness with respect to laser amplifiers, it is also an undisputable strength when it comes to handling high average power and wavelength tunability. Indeed, when OPA crystals are transparent to all three waves, there is, in principle, no absorption nor resonant effects in the gain medium (no population inversion). This statement shall actually be mitigated when one takes into account the other linear (partial absorption of the idler wave) or non-linear effects (two-photon absorption at the pump wavelength, for example), but OPCPA still outperforms most CPA laser amplifiers in the NIR or MIR when it comes to wavelength versatility and high average power.

Another key difference between OPA/OPCPA and lasers is that three-wave mixing processes are phase-sensitive. This property is exploited to produce optical sources with a reproducible electric field: a system which delivers not only pulses with the same temporal intensity profile but also with a constant phase between the optical carrier and the amplitude envelope (CEP, for carrier-envelope phase). To date, OPCPA offers the best technological solution to produce CEP-stable pulses in the NIR and MIR.

Last but not least, the temporal contrast of OPCPAs is orders of magnitude better than common femtosecond lasers; in the order of 10^{12} compared to the normal 10^{6} of a Ti:Sa laser. This is due to the high-order non-linear process that generates the seed.

The present chapter is organized as follows. In section 4.1, we describe the global architecture of a tunable high-average-power OPCPA. In section 4.2, we detail the physical processes of filamentation in bulk media. Section 4.3 focuses on parametric amplification. Several non-linear crystals are presented with their advantages and drawbacks. The last section is dedicated to dispersion management in the OPCPA chain.

4.2 Architecture

In principle, seed and pump pulses may be generated independently, by two synchronized laser systems for instance. In practice, the requirements on synchronization are so stringent that passive optical synchronization is often preferred to an active electronic one. A common scheme in the 2010s is to start from a common ultrabroadband femtosecond oscillator to seed both the OPCPA and the pump laser with an optional ancillary DFG in a bulk crystal to extend the wavelength coverage. The latest embodiment of this idea relies on cascaded sections of laser amplification (in Erbium-doped or Thulium-doped fibers), spectral broadening (non-linear fibers) and spectral selection (bandpass filters), all of which are fibered.

The current trend is, however, to generate a broadband seed directly from the (high-energy) pump pulses through filamentation of picosecond pulses in bulk crystals. This is actually the key design of compact tunable sources in the NIR and Mid-IR wavelength range. If white-light generation has been used for decades to seed OPAs pumped by short pulses at 800 nm, the filamentation regime induced by picosecond pulses in long crystal (>10 mm) exhibits quite different properties. In particular, the octave-spanning infrared wing of the generated continuum is remarkably smooth, stable and robust [14–16]. Of most interest for carrier-envelope phase stability, the filamentation process does not add a significant level of phase noise in the infrared despite the high level on non-linearity. Continuum generation from pump pulses minimizes the optical path between the reservoir of energy, i.e., the pump, and the seed pulses in the OPA stages. It limits the effects of the environment (temperature and vibrations primarily) and, consequently, limits the delay drift/jitter between the two arms as well as dispersion variations (figure 4.2).

Another challenge in high-power tunable OPCPAs is the global dispersion scheme, which may be divided into two sub-challenges. The first obstacle is the small stretching ratio of picosecond-pumped OPCPAs, typically in the range of 5–25, which forbids grating based compressors or stretchers and favors bulk media and dispersive mirrors. The second difficulty is tunability itself. It is no exaggeration to say that tunability and chirped-pulse amplification are technologically hardly compatible. A large tunability implies a large bandwidth coverage while CPA requires a constant dispersion. The result is that the stretcher and compressor should be both ultrabroadband and of constant dispersion (constant chirp), which implies no high-order phase such as third- or fourth-order phase. Dispersive mirrors are probably the only devices fulfilling this requirement. Unfortunately, the limited time-bandwidth product of dispersive mirrors (which are optically 'thin' devices) cannot accommodate both a large bandwidth and a large dispersion per bounce: the dispersive power and the reflectivity both scale inversely with the reflection bandwidth. Also, the different frequencies will not experience the same index through the same material, the group velocity dispersion (GVD) leads to different pulse duration in the OPAs in function of the seeded frequency. A solution to these dilemmas is to combine sets of dispersive mirrors for pulse compression with an adaptive pulse stretching (pulse-shaping device).

Last, the parametric amplification of nJ-level seed pulses to the mJ-level corresponds to a single-pass saturated gain of ~10^6, which is a value that exceeds what a single OPA stage can achieve in broadband operation. The parametric gain is therefore distributed among several stages (see figure 4.2), each of which are operated in the saturation regime.

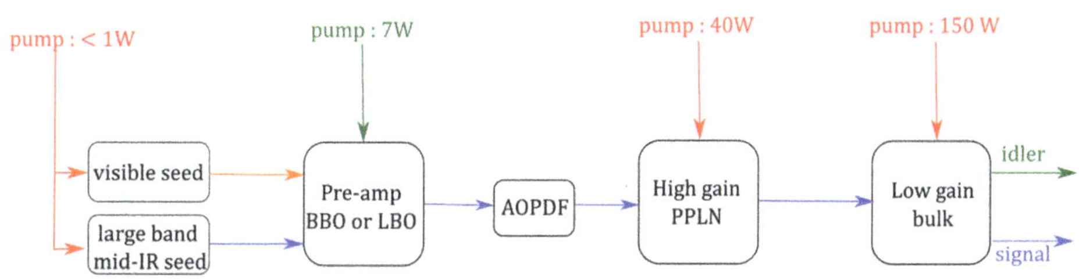

Figure 4.2. Schematic of the chain of amplification for a dual system with a tunable <100 fs pulse duration operation and a <50 fs CEP-stable pulse centered at 1.7.

As a guideline, we describe in this chapter a tunable OPCPA pumped by a Dira-200 laser from TRUMPF Scientific Lasers (2 mJ, 1 ps, 100 kHz, 1030 nm). The OPCPA is tunable from 1.4 µm to 4 µm with a gap around 2.1 µm, and the user can select to have either a CEP-stable signal at 1.75 µm or a CEP-stable idler at 3 µm. The system is built around the following stages:

- A continuum generation stage (WLG) providing a 1.4–1.85 µm seed. This stage is pumped at 1030 nm.
- An optional second-harmonic stage to produce a seed centered around 800 nm from the WLG stage.
- A first OPA/DFG stage pumped at 515 nm (BBO crystal, collinear).
- An acousto-optic programmable dispersive filter (Dazzler, Fastlite) to stretch the seed pulses, control the dispersion and the CEP.
- A first pre-amplifier pumped at 515 nm (LBO crystal, non-collinear).
- A second pre-amplifier pumped at 1030 nm (PPLN crystal, non-collinear).
- A power amplifier pumped at 1030 nm (LN crystal, collinear).
- A dichroic mirror that separates the signal and idler beams.
- A pulse compressor for the idler beam (bulk windows).
- A pulse compressor for the signal beam (dispersive mirrors).
- A beam sampler for embedded diagnostics.

Depending on the amplification stage, the design of each non-linear stage was selected to fulfill a given function: broadband amplification but with low efficiency (BBO, LBO), amplification at high efficiency (PPLN) and small-gain amplification at high power (LN).

4.3 Generation of a broadband mid-IR seed from a narrowband picosecond pulse

In this section, we will explain the different processes involved in the generation of a broadband continuum during filamentation in a bulk medium with a picosecond pulse. The purpose here is to give a general picture and physical understanding, not a detailed description of these phenomena. A detailed description of filamentation in solids can be found elsewhere [17].

Filamentation is a very rich process involving several non-linear and linear phenomena. This complexity makes it very difficult to predict theoretically the possible continuum spectrum. Therefore, experimental spectra for the experimental parameters considered in this chapter will be shown.

4.3.1 Optical Kerr effect: self-focusing and self-phase modulation

When an electric field is applied to a medium, the latter will react and its reaction can be described by a vector called the polarization vector (4.1). The full knowledge of the polarization response of the medium involves microscopic understanding of the solid. However, the macroscopic response can be expanded to a linear term and several non-linear coefficients. The linear part is well known and corresponds to the refractive index of the medium. It depends on the electric field $E(t)$ and the linear susceptibility $\chi^{(1)}$. The non-linear terms will be approximated in this chapter to only the first two terms. Due to the symmetry of the medium in the case of a non-birefringent crystal (i.e., assuming a centrosymmetric crystal), the quadratic part of the non-linear polarization vanishes.

$$\boldsymbol{P}(t) = \epsilon_0(\chi^{(1)}\boldsymbol{E}(t) + \chi^{(2)}\boldsymbol{E}^2(t) + \chi^{(3)}\boldsymbol{E}^3(t), + ,\dots) \tag{4.1}$$

$P(t)$ is the polarization vector, $E(t)$ is the electric field, ϵ_0 is the absolute permittivity and $\chi^{(i)}$ are the linear and non-linear susceptibilities.

The total polarization vector will consists of the linear term plus the non-linear response, function of $\chi^{(3)}$ and the electric field multiplied by the intensity. Finally, for a linear polarization, this will be equivalent to a refractive index containing its linear part and a non-linear component, function of the pulse intensity and the non-linear refractive index n_2I (4.2)

$$n = n_0 + n_2 I \tag{4.2}$$

n_0 being the linear refractive index and I the laser intensity.

In this section, we will focus first on the non-linear response through its non-linear refractive index n_2I, named the Kerr effect. This will trigger two main processes, self-focusing and self-phase modulation. Both phenomena result from acquiring an intensity dependent phase term during propagation described by (4.3).

$$\Phi_{nl}(x, y, t) = \frac{\omega_0}{c} n_2 \int I(x, y, z, t)dz \tag{4.3}$$

Φ_{nl} is the non-linear phase, ω_0 is the central frequency and c the light velocity in vacuum.

Let us focus first on the spatial dependence of the linear refractive index. Indeed, since the laser intensity is spatially inhomogeneous for a Gaussian beam, being higher in the center, the spatial phase acquired is higher in the center of the beam and will result, in media characterized by a positive n_2, in a self-focusing effect. The critical power at which the self-focusing effect counter-balances the self-diffraction is given by (4.4) [18].

$$P_{cr} = \frac{\alpha \lambda^2}{4\pi n_0 n_2} \quad \text{with} \quad \alpha = \frac{1.22\pi^2}{8} \tag{4.4}$$

P_{cr} is the critical power, λ is the central wavelength.

As described in figure 4.3, at a certain point of propagation in the medium, the beam will collapse. This effect will not lead to a singularity because ionization of the medium will counter-balance the self focusing effect through free electrons, as will be explained in the next section.

The non-linear phase term (4.3) is also time-dependent. This effect, called self-phase modulation, leads to the generation of new frequencies, in other words, to spectral broadening. The spectral phase due to self-phase modulation shows a non-trivial profile but can be approximated to a positive chirp, meaning lower frequencies are created at the leading edge of the pulse and higher frequencies at

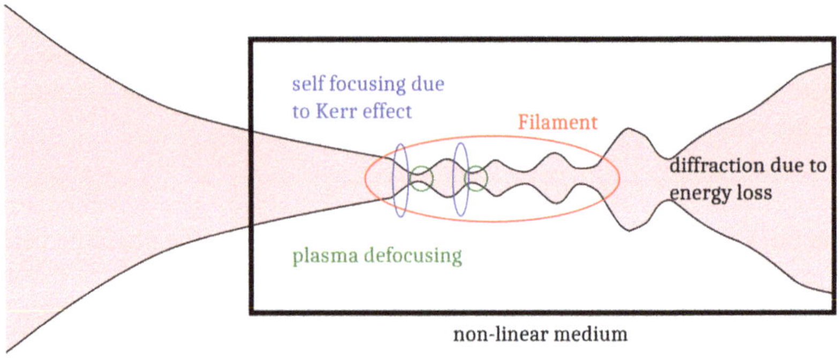

Figure 4.3. Qualitative picture of filamentation in a bulk non-linear medium. Adapted in part with permission from [43] © The Optical Society.

the trailing edge. This is the main effect responsible for the generation of a broadband week seed from a narrowband intense picosecond pulse. However, for some conditions where a soliton can propagate, delayed Raman effects can play a significant role in transferring energy from higher to lower frequencies [19, 20]. In this case, the non-linear phase will take the form:

$$\Phi_{nl}(x, y, z, t) = \left\{ \frac{\omega_0}{c}(1 - \alpha)n_2 I(x, y, z, t) \right\} + \left\{ \alpha \frac{\omega_0}{c} n_2 \int R(t - \tau) I(x, y, z, t) d\tau \right\} \tag{4.5}$$

with $R(t) = \alpha H(t)\Omega \exp(-\Gamma t)\sin(\Lambda t)$, $\Omega = \frac{\Lambda^2 + \Gamma^2}{\Lambda}$ [21].

Λ and Γ are characteristic frequencies associated with the delayed Raman response of the medium, α is the ratio of each process and $H(t)$ is a Heaviside step function. In equation (4.5), the first term of the sum corresponds to the 'classical' instantaneous Kerr effect, and the right term corresponds to the delayed Raman effect. Typical Raman response time in solids spans from 100 fs to a few picoseconds.

4.3.2 Plasma generation: multi-photon ionization

When a laser pulse propagates through a non-linear medium, the latter will experience a self-focusing effect as explained before. This will induce an increase of the peak intensity and lead to multi-photon absorption. This is a non-linear process that allows electrons from the valence band to be excited to the conduction band by absorption of several photons. Its probability will be proportional to the laser intensity to the power of n, I^n, n being the number of photons involved in the process.

The generated free electrons will induce a change in the refractive index and will also contribute to an additional phase term which can be described [17] by

$$\Phi_{nl}(\boldsymbol{x}, t) = -\frac{\omega_0 L}{c} \frac{\rho(\boldsymbol{x}, t)}{2 n_0 \rho_c}$$

where Φ_{nl} is the non-linear phase due to free electrons, L the length of the medium, $\rho(x, t)$ the density of free electrons and $\rho_c = \frac{\epsilon_0 m_e \omega_0^2}{e^2}$ is the critical density at which the medium becomes non-transparent. m_e and e are, respectively, the electron mass and charge.

Since the electron density is a positive quantity, the non-linear phase due to free electrons will be negative and will therefore counter-balance the effect of self-focusing due to the Kerr non-linearity. The successive focusing due to the Kerr non-linearity followed by the defocusing due to the ionized electrons is responsible for the formation of a filament. Eventually, due to absorption in the medium, the laser intensity will decrease and the self-diffraction will become the main physical process leading to the defocusing of the beam.

4.3.3 Linear dispersion and other effects

The refractive index not only depends on pulse intensity but also has a linear part, which depends on the wavelength. While for a narrowband, this wavelength dependence may not matter. While new frequencies will be generated in the medium, their spectral phase will be affected through propagation in the medium. Most of the materials used to generate a continuum by filamentation have a normal (positive) GVD in the visible/NIR range and an abnormal (negative) dispersion in the NIR-mid-IR. It means that for the wavelength range of interest in this chapter, the linear dispersion of the medium will be negative. Since the spectral phase originated from self-phase modulation has a positive second order, these effects will compensate each other, and it is often considered that the NIR part of the continuum is close to the Fourier-transform-limited duration.

The Kerr effect not only shows non-linear dependence of the refractive index but also to the group velocity through non-linear dependence of the group index. The latter will be higher for higher intensities, i.e., the group velocity will be lower at the center at the pulse. This results in a temporal steepening of the trailing edge and a smoothing of the leading edge.

4.3.4 Experimental results

In practice, all these phenomena are extremely difficult to model precisely. Experimental trials are often needed in order to find the right medium for a specific application. In table 4.1, we show the non-linear index values and bandgap energies for several common materials. Qualitatively, the material should have a high non-linear index with a high bandgap energy, resulting in a strong broadening due to self-phase modulation, while multi-photon ionization would require very high intensities, allowing one to use higher intensities. In practice, the quality and damage threshold [22] of the crystals is of high importance. Color centers due to defects in the crystals, showing lower local bandgap energies, will induce more electrons in the conduction band. This will have an influence on the instantaneous response of the medium but also will build up over several laser shots when the repetition rate becomes high enough.

Table 4.1. Non-linear index and energy bandgaps of common materials used for continuum generation from a 1030 nm picosecond pulse.

Material	Non-linear index n_2 (10^{-16} cm^2 W^{-1})	Energy bandgap (eV)
Sapphire	3.06 [23]	9.9 [24]
YAG	6.13 [25]	6.3 [26]
Fused silica	2.19 [27]	8.4 [24]
CaF_2	1.71 [27]	10 [24]

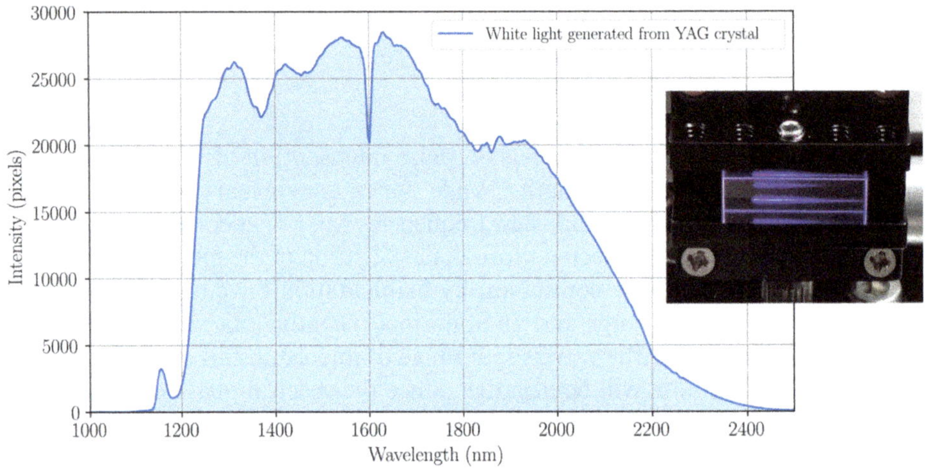

Figure 4.4. Continuum spectrum generated from a picosecond pulse in a 15 mm long YAG crystal, optimized for bandwidth. The spectrum is recorded after a long pass filter selecting the wavelength above 1.2 μm. The blue color of the filament on the picture corresponds to the fluorescence of excited states populated by multi-photon absorption in the crystal. Adapted in part with permission from [44] © The Optical Society.

In the practical case of generating a broadband NIR seed from a picosecond pulse centered at 1030 nm, the YAG crystal shows the best results [23]. However, as soon as the repetition rate exceeds several kilohertz, the medium has to be regularly refreshed in order to avoid a build up of ionization or damage.

Figure 4.4 shows an experimental continuum spectrum generated from a picosecond pulse centered at 1030 nm wavelength in a 15 mm YAG crystal. By optimizing the filamentation conditions, it is possible to extend the spectrum towards longer wavelength up to above 2.2 μm.

4.3.5 Generation of visible seed

All the previous sections have considered the generation of a broadband NIR seed in order to amplify selectively a certain part of the seed spectrum to generate a high-power tunable NIR-mid-IR source. As explained in figure 4.1, for each amplified seed, an idler will be generated. As we will explain later, this will lead to a tunable source from 1.4 μm to 1.85 μm central wavelength for the signal beam and from

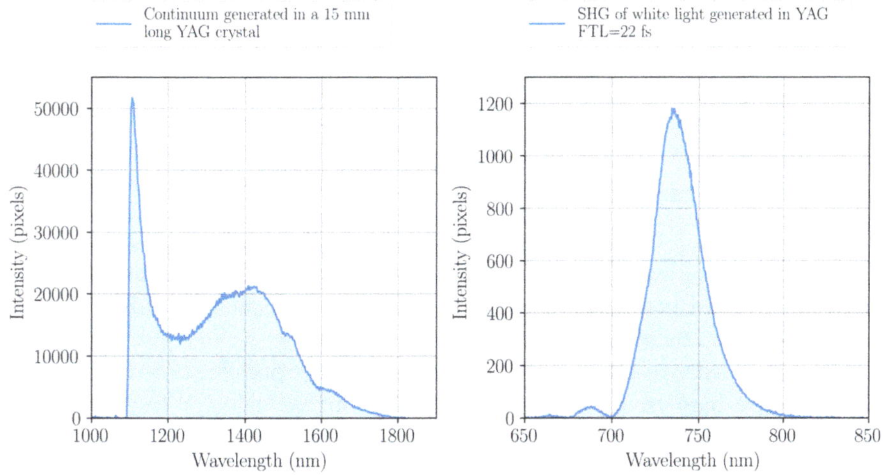

Figure 4.5. Left, continuum generated in a 15 mm YAG crystal, optimized for the highest flux around 1.5 µm wavelength. Right, SHG spectrum of left continuum from a 2 mm BBO.

2.3 µm to 3.9 µm for the idler beam. However, due to phase difference during the amplification process, only the idler part will have passive stable CEP, meaning the phase of the electric field will not fluctuate pulse-to-pulse. The signal beam will keep the same phase than the pump pulse, which is random. However, if one wants to generate a CEP-stable seed in the near-IR wavelength range covering the signal spectrum, it must be generated through the difference frequency generation (DFG) between the pump and a visible seed (see section 4.4.2 for more explanation).

The approach we propose in this chapter is the second-harmonic generation (SHG) from the NIR seed generated from filamentation in a YAG crystal. Indeed, even with low intensity, it is possible to efficiently generate the second harmonic from the NIR part of the continuum in a BBO crystal (figure 4.5). In order to generate a NIR pulse with a central wavelength centered around 1.75 µm, one needs to make a DFG between the pump centered at 1030 nm and a visible seed centered around 740 nm.

4.4 Design of a tunable amplification chain

The following parts show a possible scheme to amplify in the same chain a tunable NIR seed without CEP and a broadband NIR CEP-stable seed (see figure 4.2). It gives a constraint on the first pre-amplifier. Indeed, the latter will be seeded from two different continuum sources. The first one will range from 1.4 µm to 1.85 µm central wavelength. The second one is a visible pulse from which a CEP-stable, 1.7 µm pulse will be generated through a DFG process.

The next sections will show the different options in order to amplify the NIR seed up to several hundreds of microjoules at 100 kHz repetition rate when pumped with 200 W. As will be explained in chapter 5, due to dispersion management, only the signal beam ranging from 1.4 µm to 1.85 µm is amplified until the last amplification stage where signal and idler will be compressed.

The purpose of this chapter is not to explain in detail the principle of parametric amplification. However, certain concepts will be used and we will briefly introduce them. For a detailed description of non-linear processes and parametric amplification, we refer the reader to ref. [28].

4.4.1 Concept of parametric amplification

While in the case of a centrosymmetric medium, the non-linear susceptibility of the second order $\chi^{(2)}$ can be neglected, in the present section, we will consider the case of a non-zero value. Even if other non-linearity like self-phase modulation is still possible, it will be neglected.

The particular case of parametric amplification is the possibility to transfer energy from an intense high-frequency pulse, called pump, to a lower frequency weaker seed. This effect will mainly be dependent on the value of $\chi^{(2)}$, given by the medium, and the matching of the phase velocities of all the waves involved, called phase matching.

The $\chi^{(2)}$ quantity is a tensor, dependent on the three spatial coordinates. However, for a particular orientation of the non-linear medium with regard to the incident light wave-vector, it can be reduced to a scalar quantity called effective non-linearity d_{eff}. As in the case of the non-linear refractive index, the larger the effective non-linearity d_{eff}, the larger the non-linear parametric effect.

Another important requirement to be fulfilled is the necessity to match the phase velocities of all waves involved in the process so they can interact constructively (see figure 4.6). This can be mathematically expressed as

$$\Delta \boldsymbol{k} = \boldsymbol{k}_{\text{pump}} + \boldsymbol{k}_{\text{signal}} + \boldsymbol{k}_{\text{idler}} = 0 \tag{4.6}$$

with $\boldsymbol{k}_x$ being the wave-vector defined by $\frac{2\pi n_x}{\lambda}$, n_x being the refractive index of the considered pulse. This condition can be fulfilled in different ways.

First of all, the use of birefringence, where the refractive index of the medium is dependent on the polarization of the incident wave, can provide such conditions. Indeed, in a birefringent crystal, it is possible to define two perpendicular axes in the plan of polarization orthogonal to the direction of propagation. A wave propagating through such a birefringent crystal will experience a different refractive index if its polarization is collinear to the ordinary or the extraordinary axis. The combinations

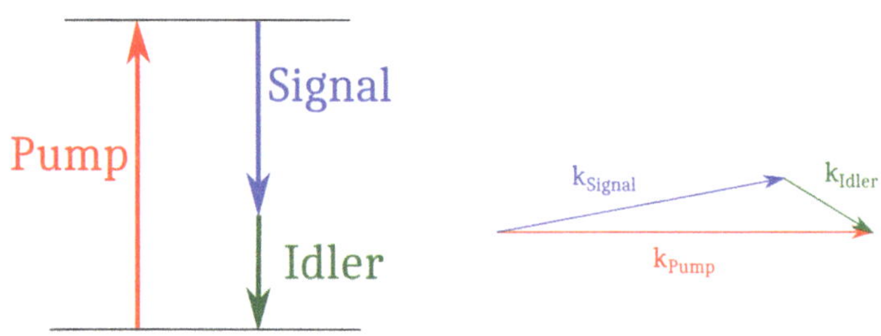

Figure 4.6. Physical picture of parametric processes in a non-linear medium. The image on the left illustrates the conservation of energy while the image on the right shows the conservation of momentum.

of polarization have to fulfill the conditions of equation (4.6). They will be divided in three types:

- Type 0: pump, signal and idler have same polarization.
- Type I: signal and idler waves have same polarization, pump polarization is different.
- Type II: signal and idler have crossed polarization.

The other phase-matching type, called quasi-phase matching, is the tailoring of the non-linear susceptibility as a function of propagation length. Indeed, by inverting the effective non-linearity periodically with a period Δ, it is possible to constructively build up the second-order non-linearity over propagation. Each part of the non-linear crystal with a constant sign of d_{eff} is called a domain. This is modeled by adding to the phase mismatch equation (4.6) an additional wave-vector originating from the periodic inversion of the d_{eff},

$$\Delta \boldsymbol{k} = \boldsymbol{k}_{\mathrm{pump}} + \boldsymbol{k}_{\mathrm{signal}} + \boldsymbol{k}_{\mathrm{idler}} + K = 0$$

where $K = \frac{2\pi}{\Delta}$. Δ corresponds to two times the domain width. It will be used with a type 0 phase matching.

4.4.2 Pre-amplifier for a tunable source and generation of a CEP-stable seed

As explained before, the pre-amplifier must be compatible with the DFG between a visible seed and the pump as well as the amplification of the NIR seed. The requirement on the pump wavelength to be shorter than the seed and idler enforces the use of the SHG of the 1030 nm beam centered at 515 nm. It can be achieved with a type I BBO crystal, with a thickness chosen either to saturate the SHG process in order to get a stable 515 nm beam, or to tune the conversion efficiency to the desired value in order to re-use the non-converted pump. Indeed, when the SHG process is saturated, the 1030 nm pulse is depleted strongly in its temporal and spatial center making it undesirable for future use.

In the case of a tunable light source, most applications will not benefit from short temporal pulses since they are used to excite a system resonantly. For example, the coherent build up of a population in an excited state requires several optical cycles to be efficient. Moreover, if one wants to operate the system with homogeneous performances over the whole bandwidth, the desired pulse duration is imposed by the longest one. Indeed, due to spectral dependent phase matching, the amplification bandwidth is not homogeneous over the total bandwidth. The chosen pulse duration for the tunable operation for such an OPCPA system must be around 100 fs.

The broadband capabilities of BBO or LBO crystals pumped at 515 nm with a type I phase matching make them suitable choices for the pre-amplifier. Figure 4.7 shows the amplified seed from a broadband continuum generation in a YAG crystal. The dispersion of the seed is adjusted in order to amplify only the desired bandwidth of the seed spectrum, corresponding to a <100 fs pulse. The complementary wavelengths span here from 710 nm to 810 nm.

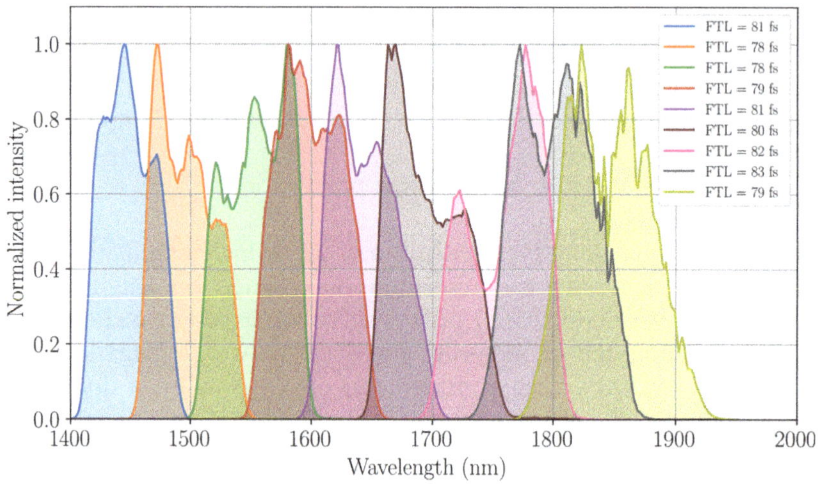

Figure 4.7. Experimental spectra from the amplification by a 515 nm pump of a broadband continuum generated in a YAG crystal. The crystal phase-matching angle is changed in order to select which spectral part is amplified. Seed chirp is adjusted for below 100 fs operation, so delay between pump and seed is adjusted for each wavelength as well. FTL: Fourier-transform limit.

As shown in figure 4.2, the first pre-amplifier is seeded from two different sources. The previous one was the broadband NIR continuum. Nevertheless, in order to generate a passively CEP-stable short NIR pulse, the same pre-amplifier needs to be able to achieve DFG in the same crystal. Indeed, during the DFG process, the complementary beam has a frequency equal to the difference between the pump and seed,

$$w_c = w_p - w_s,$$

w_c, w_p and w_s are, respectively, the complementary, pump and seed angular frequency.

The CEP Φ_c will also show the same relation,

$$\Phi_c = \Phi_p - \Phi_s.$$

If Φ_p and Φ_s share a phase relation which is not fluctuating shot-to-shot, the phase difference Φ_c will have a constant value. The complementary pulse is called CEP-stable.

4.4.3 Periodically poled crystals for high amplification stages

The pre-amplifier allows one to reach the microjoule level from a sub-nanojoule pulse. However, in order to reach the maximum output, it is still needed to apply a total gain of several hundred. It is not possible to achieve this value from a single amplification stage while maintaining the bandwidth covering the tunability range.

A very interesting choice for a high gain amplification stage is the use of a quasi-phase-matching crystal (Type 0), the periodically poled lithium niobate (PPLN) [29]. As explained in section 4.4.1, the periodic structure of the non-linear susceptibility

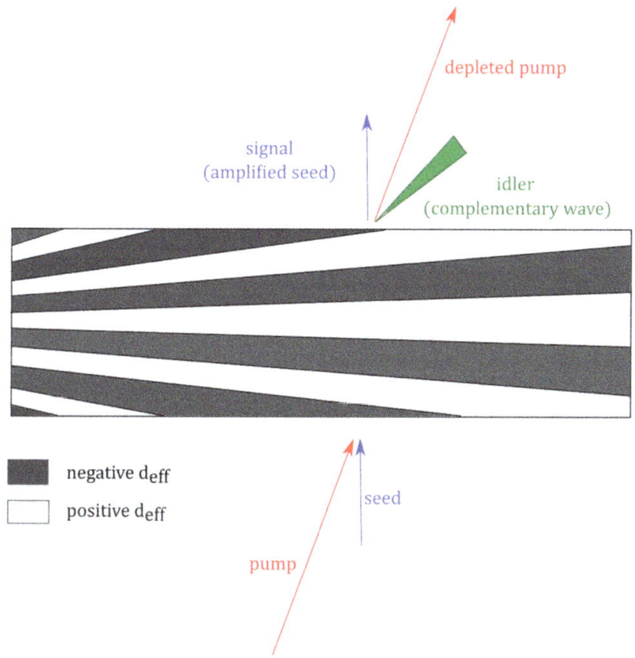

Figure 4.8. Periodically poled crystal with a fan-out geometry. The period of the poling is dependent on the coordinate perpendicular to the beam propagation direction.

allows very efficient interaction between the pump and the seed. The effective non-linearity is approximately 15 pm V^{-1} while for a bulk lithium niobate (LN) crystal, this value is equal to 4 pm V^{-1} (at phase matching). However, the poling period is specific to a particular central wavelength with a certain bandwidth. Unfortunately, this bandwidth is generally not enough to amplify the full tunable range. It is necessary to change the poling period for each central wavelength of the tuning range. While it would be possible to design a PPLN crystal with multiple periods, a more suitable approach is the use of a fan-out geometry [30] (see figure 4.8). By using this type of phase-matching geometry, it is possible to tune the phase-matched wavelength by translating the PPLN crystal perpendicularly to the beam propagation direction.

Because of the presence of an acousto-optic programmable dispersive filter (AOPDF) between the pre-amplifier and the PPLN amplification stage, only 10% of the total energy from the pre-amplifier is available for seeding the PPLN high gain stage. Gain as high as 50–100 is obtained depending on the central wavelength. The output of this stage spans from 0.5 W to 3 W for a bandwidth spanning from 1.45 μm to 1.85 μm central wavelength (see figure 4.9).

4.4.4 Low gain amplification stage

Unlike the previous amplification stage, the last low gain stage will have an all-collinear geometry. Indeed, in the case of a non-collinear geometry, due to

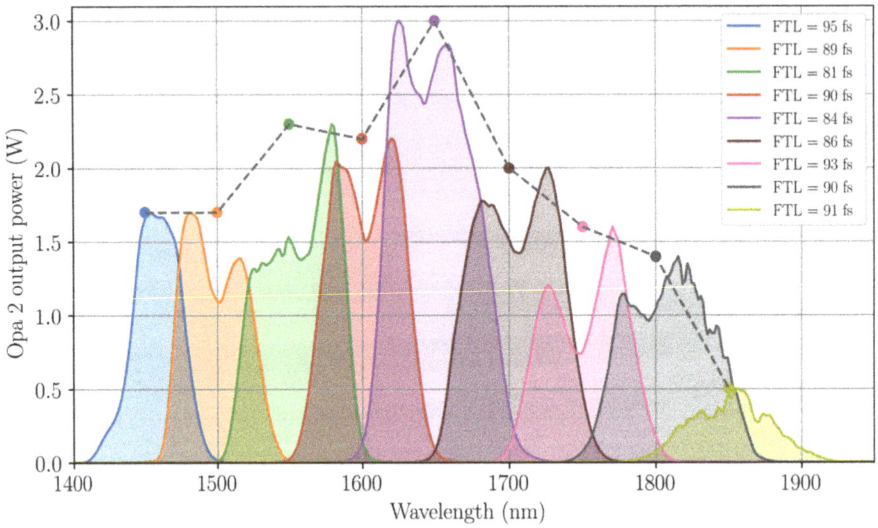

Figure 4.9. Amplified spectra from the amplification stage based on a 1-mm-thick fan-out PPLN crystal, pumped with 40 W. All spectra are compatible with sub-100 fs pulses.

momentum conservation, the complementary beam sees its wave-vector having a wavelength dependent emission angle, called also spatial chirp. It is not possible to use this beam directly after amplification. Therefore, the use of an all-collinear geometry is chosen (see figure 4.8).

The purpose of this amplification stage is to extract a maximum energy from the pump and transfer it to an already intense seed. Therefore, the gain will not be very high, between 5 and 10. A fan-out PPLN crystal is not a suitable choice for this last stage since the pump beam will have a rather large diameter into the non-linear crystal in order to adjust the desired intensity. PPLN crystals cannot be made with arbitrary size and will be too small for this stage. Moreover, it will not benefit from the very highly effective non-linearity since only low gain values are targeted. This very high non-linear susceptibility can even lead to efficient generation of parasitic wavelength. They may be absorbed in the non-linear crystal and, in the case of very high average power, can lead to damage of the crystal.

It is then suitable to use a bulk crystal with a type I or type II phase-matching scheme using birefringence. Several candidates are possible. We present here the case of a 2-mm-thick LN crystal used in type I phase matching. Its effective non-linearity is approximately equal to 4 pm V^{-1} over all the bandwidth. Its phase-matching bandwidth spans from a 95 fs FTL pulse at 1.4 μm central wavelength (3.9 μm complementary) to below 40 fs when getting closer to degeneracy.

Figure 4.10 shows the signal and idler output spectra of the OPCPA system for different central wavelengths. The total tunable bandwidth ranges from <1.45 μm to almost 4 μm, with a gap around 2 μm at the degeneracy.

Due to very high intensities and average power, this last amplification stage suffers from several parasitic effects. As can be seen in figure 4.10, absorption bands

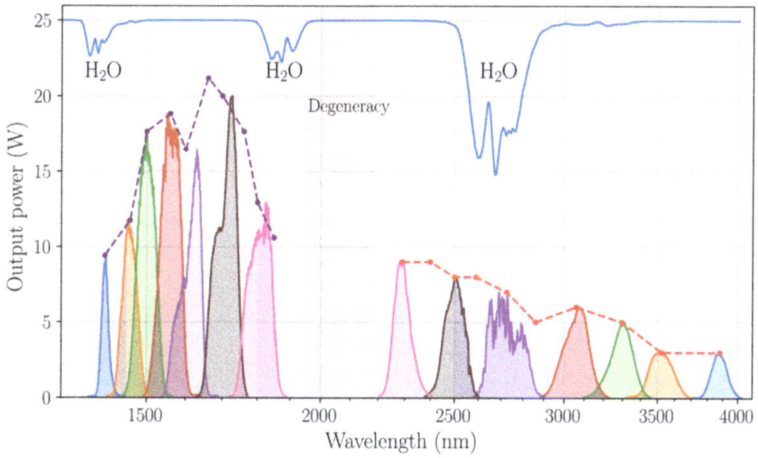

Figure 4.10. OPCPA output spectra for signal (left part below 2000 nm) and idler (right part above 2000 nm) for different central wavelengths generated in a 2-mm-thick LN crystal pumped with approximately 140 W. Spectra are normalized to the output power. It ranges from 10 W to 20 W for the signal beam and from 3 W to 9 W for the idler beam. FTL pulse durations are between 95 fs and 105 fs. The top curve shows the absorption spectrum of water vapor.

of water vapor overlap with the output wavelength range. In order to avoid absorption from water in air, leading to decreased power and longer pulse duration, it is necessary to operate the system under dry air or nitrogen. Absorption in air from water molecules can also lead to a degradation of the spatial properties of the beam [31]. The LN crystal needs also to be heated to high temperature (around 120 degrees Celsius) in order to lower the photo-refractive effects [32, 33]. Finally, since the phase-matching angle of the LN crystal is dependent on the crystal temperature [34–36], it is important to wait for full temperature equilibrium to set optimal parameters for a specific wavelength. This can be achieved rather quickly regarding heating due the crystal optical absorption (within a few minutes).

4.4.5 Compatibility with a CEP-stable output for strong field applications

In section 4.4.2, we explained the possibility in a single stage, seeded from two different sources, to generate two different outputs: a tunable one and a broadband CEP-stable pulse. Both are amplified in the same amplification chain. However, the dispersion of the CEP-stable pulse has to be adjusted in a manner that it is not compatible with tunability. As will be seen in section 4.5.1, the spectral phase shaping, in order to efficiently amplify all the different wavelengths of the tunable mode and the broadband short pulse, is done by the use of an AOPDF. It is a key device in the amplification chain.

Figure 4.11 shows the output spectrum from the same amplification chain than the tunable mode but with a different seed. It is compatible with a 48 fs pulse duration with an output power of 19.3 W at a 100 kHz repetition rate. The main reason for a different seed is the necessity to realize a DFG process leading to CEP stability (see figure 4.12).

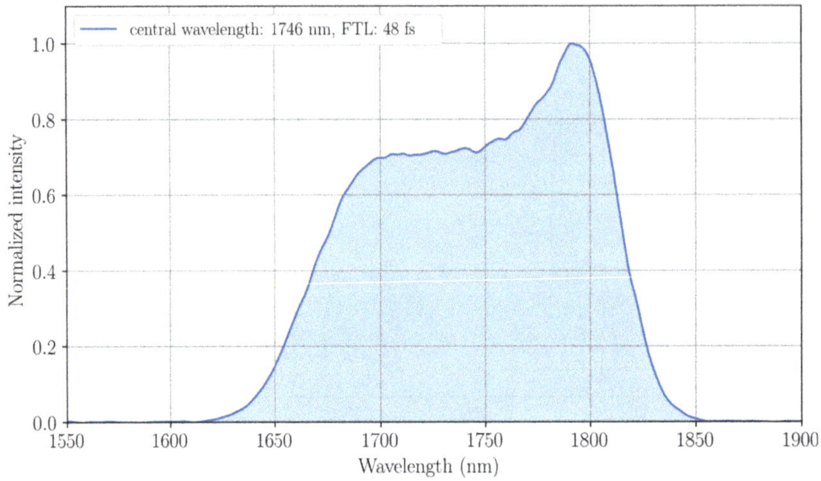

Figure 4.11. Output spectrum for the short pulse mode with CEP stability. Output power is 19.3 W at 100 kHz repetition rate after compression.

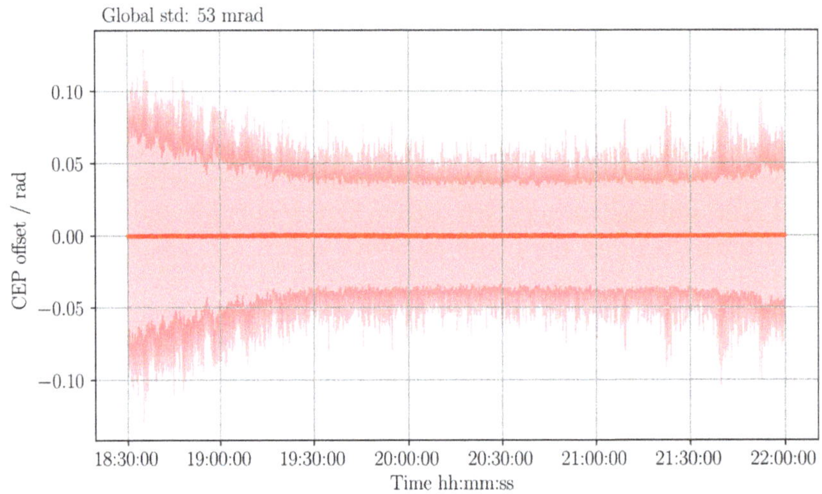

Figure 4.12. CEP measurement from a f-2f setup with a Fringeezz device. Single shot CEP noise as low as 53 mrad over a period of 3 h 30 min can be achieved. The thick red line is the average value of the CEP over 1 s (100 000 shots) and the filled area around the average value shows the standard deviation of the CEP values over 1 s.

4.5 Dispersion management and compression

4.5.1 The use of an acousto-optic programmable dispersive filter

An acousto-optic programmable filter (Dazzler device [37, 38]) allows direct shaping of the spectral phase and amplitude of a pulse. It is a key device in tunable OPCPA systems. Indeed, as can be seen in figure 4.13, most of the materials used for lenses and non-linear crystals show a wavelength dependence of the second- and third-

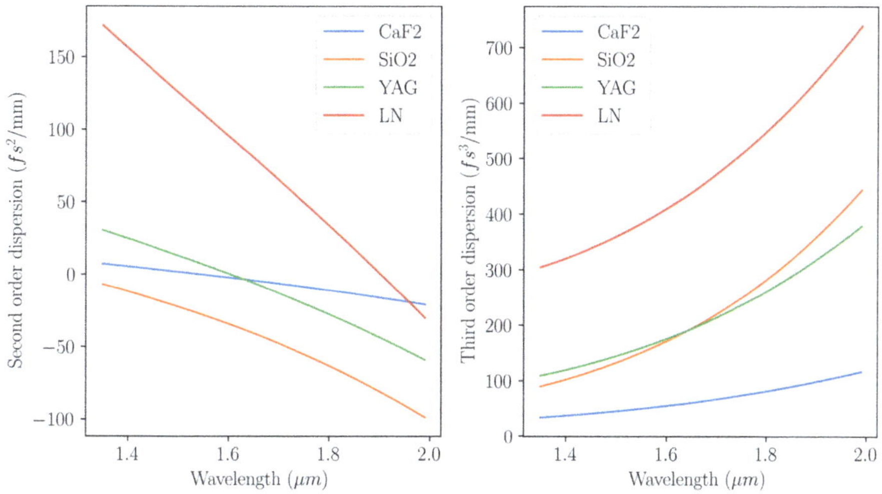

Figure 4.13. Left, group velocity dispersion (GVD) of several media. Right, third-order dispersion (TOD) of several media. Sellmeier equations were obtained from [39–42].

order dispersion. For a fixed central wavelength, it is possible to design a compressing setup correcting those two terms but in the case of a tunable system, the only possibility to get a nearly Fourier-transform-limited temporal pulse at the output of the system is to dynamically change the dispersion.

Various other advantages exist from using an AOPDF. For example, it is possible to drastically reduce the amplified bandwidth in order to scan very precisely through a resonant process. If the resonance scan is done within the bandwidth of the <100 fs mode, no other mechanical part has to be moved. Therefore, the central wavelength can be changed shot-to-shot at a 100 kHz repetition rate.

Finally, the other key role of the device is to allow easy switching between the tunable mode and the short CEP-stable pulse mode. Since the bandwidths are not the same, the dispersion needed for matching temporally the pump and seed in the different amplification stages is different.

4.5.2 Compression setup for a multi-mode OPCPA system

The final step of an OPCPA system is to compensate for chirp added to the pulse. However, designing a compression scheme for a tunable system for this extremely large bandwidth is a complex task.

The system has to be divided into three different parts. First, the signal and idler beams cannot be compressed with the same device. Indeed, their wavelengths are very different but mostly, their chirp is of opposite sign due to the parametric process. In the case of the signal beam, the tunable mode and the short CEP mode also need separate compressors due to very different values of the chirp.

One of the major constraints is the very broad bandwidth covered by the idler beam. The usual methods for compressing pulses like a grating based compressor or a set of chirped mirrors are not applicable because no manufacturer can make optics

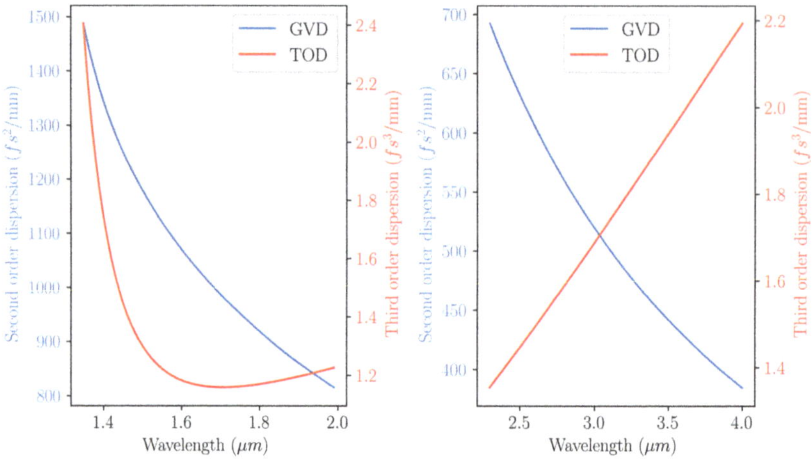

Figure 4.14. Left, GVD and TOD of silicon for the signal wavelength range. Right, the same quantities for the idler wavelength range. The Sellmeier equation for silicon is obtained from [42].

dispersive homogenously over such a broad range. Therefore, the choice for compression of the idler beam is to use a bulk compressor. The ideal material in order to keep the compressor on reasonable thicknesses and regarding the ratio GVD-TOD is silicon. The GVD and TOD curves for silicon over the two bandwidths of the signal and idler beam are shown in figure 4.14. Regarding the positive value of the GVD of silicon over the idler bandwidth, it is necessary to generate an idler beam with a negative chirp. A consequence of this is a positive chirp for the signal beam in the amplification stages. The AOPDF device again plays an important role in order to compensate the third order from the bulk compressor.

In order to negatively chirp the signal beam, it is difficult to use the usual glass since most show very small GVD or even change the sign of GVD in the signal tunable bandwidth. The AOPDF allows one to arbitrarily choose the negative chirp applied to the signal beam coming from the pre-amplifier.

The most suitable choice for compressing the signal beam is to use sets of chirped mirrors because they have very good reflectivity and are easy to use. However, only one set of chirp mirrors cannot cover the total bandwidth of the tunable signal while compensating for the total chirp. It is then mandatory to split the compressor for the tunable signal into two sets, each covering half of the bandwidth. In the case of the short CEP mode, only one set is sufficient.

4.6 Conclusion

In this chapter, high-power tunable sources based on OPCPA technology were presented. It was shown that new picosecond light sources based on Ytterbium: YAG are suitable pump lasers for subsequent wavelength conversion. Indeed, they can provide several millijoules per pulse at a high repetition rate.

In section 4.1 we explained the physical phenomena involved in the generation of a continuum from a picosecond narrowband pulse. It is the key step in building such a tunable OPCPA system. Then, the following amplification stages were explained, each of them being designed in order to apply the required gain. The use of an AOPDF as a dynamically adjustable dispersive element helps the system to provide optimal performances over the total bandwidth. It is also possible to generate in the same amplification chain a short pulse centered around 1.7 μm central wavelength with extreme CEP stability. Finally, different compression schemes have been used regarding the requirements for each mode of the system. Again, the AOPDF device plays an important role to compensate for the different phase terms acquired during propagation in the system.

These types of sources can provide high peak intensities at a very high repetition rate while showing tunability. Their versatility makes them ideal tools for any strong field process that experiences resonances. Indeed, terahertz generation, angle-resolved photoelectron spectroscopy, time-resolved absorption spectroscopy or even high harmonic generation from solids are already benefiting from existing tunable sources. However, the very high repetition rate will lead to an increase in signal-to-noise ratio, opening up new possibilities.

References

[1] Nubbemeyer T *et al* 2017 1 kW, 200 mJ picosecond thin-disk laser system *Opt. Lett.* **42** 1381

[2] Schmidt B E, Hage A, Mans T, Légaré F and Worner H J 2017 Highly stable, 54 mJ Yb-InnoSlab laser platform at 0.5 kW average power *Opt. Express* **25** 17549–55

[3] Zapata L E, Reichert F, Hemmer M and Kärtner F X 2016 250 W average power, 100 kHz repetition rate cryogenic Yb:YAG amplifier for OPCPA pumping *Opt. Lett.* **41** 492

[4] Müller M *et al* 2016 1 kW 1 mJ eight-channel ultrafast fiber laser *Opt. Lett.* **41** 3439

[5] Zapata L E *et al* 2015 Cryogenic Yb:YAG composite-thin-disk for high energy and average power amplifiers *Opt. Lett.* **40** 2610

[6] Metzger T *et al* 2009 High-repetition-rate picosecond pump laser based on a Yb:YAG disk amplifier for optical parametric amplification *Opt. Lett.* **34** 2123

[7] Schulz M *et al* 2011 Yb:YAG Innoslab amplifier: efficient high repetition rate subpicosecond pumping system for optical parametric chirped pulse amplification *Opt. Lett.* **36** 2456

[8] Fattahi H *et al* 2016 High-power, 1-ps, all-Yb:YAG thin-disk regenerative amplifier *Opt. Lett.* **41** 1126

[9] Beetar J E, Rivas F, Gholam-Mirzaei S, Liu Y and Chini M 2019 Hollow-core fiber compression of a commercial Yb:KGW laser amplifier *J. Opt. Soc. Am. B* **36** A33

[10] Schenkel B *et al* 2003 Generation of 38-fs pulses from adaptive compression of a cascaded hollow fiber supercontinuum *Opt. Lett.* **28** 1987

[11] Budriūnas R *et al* 2017 53 W average power CEP-stabilized OPCPA system delivering 55 TW few cycle pulses at 1 kHz repetition rate *Opt. Express* **25** 5797

[12] Thiré N *et al* 2018 Highly stable, 15 W, few-cycle, 65 mrad CEP-noise mid-IR OPCPA for statistical physics *Opt. Express* **26** 26907

[13] Thai A, Hemmer M, Bates P K, Chalus O and Biegert J 2011 Sub-250-mrad, passively carrier–envelope-phase-stable mid-infrared OPCPA source at high repetition rate *Opt. Lett.* **36** 3918

[14] Špaček A *et al* 2020 Stability mechanism of picosecond supercontinuum in YAG *Opt. Express* **28** 20205

[15] van de Walle A *et al* 2015 Spectral and spatial full-bandwidth correlation analysis of bulk-generated supercontinuum in the mid-infrared *Opt. Lett.* **40** 673
[16] Indra L *et al* 2017 Picosecond pulse generated supercontinuum as a stable seed for OPCPA *Opt. Lett.* **42** 843
[17] Dubietis A and Couairon A 2019 *Ultrafast Supercontinuum Generation in Transparent Solid-State Media* (Cham: Springer)
[18] Fibich G and Gaeta A L 2000 Critical power for self-focusing in bulk media and in hollow waveguides *Opt. Lett.* **25** 335
[19] Mitschke F M and Mollenauer L F 1986 Discovery of the soliton self-frequency shift *Opt. Lett.* **11** 659
[20] Gordon J P 1986 Theory of the soliton self-frequency shift *Opt. Lett.* **11** 662
[21] Zheltikov A M 2009 Understanding the nonlinear phase and frequency shift of an ultrashort light pulse induced by an inertial third-order optical nonlinearity *Phys. Rev. A: At. Mol. Opt. Phys.* **79** 023823
[22] Gallais L and Commandré M 2014 Laser-induced damage thresholds of bulk and coating optical materials at 1030 nm, 500 fs *Appl. Opt.* **53** A186
[23] Calendron A-L, Çankaya H, Cirmi G and Kärtner F X 2015 White-light generation with sub-ps pulses *Opt. Express* **23** 13866
[24] Weber M J 2018 *Handbook of Optical Materials* (Boca Raton, FL: CRC Press)
[25] Dubietis A, Tamošauskas G, Šuminas R, Jukna V and Couairon A 2017 Femtosekundinio superkontinuumo generavimas plačios apertūros kondensuotose terpėse *Lith. J. Phys.* **57** 113–57
[26] Xu Y and Ching W Y 1999 Electronic structure of yttrium aluminum garnet (formula presented) *Phys. Rev. B: Condens. Matter Mater. Phys.* **59** 10530–5
[27] Kabaciński P, Kardaś T M, Stepanenko Y and Radzewicz C 2019 Nonlinear refractive index measurement by SPM-induced phase regression *Opt. Express* **27** 11018
[28] Boyd R W 2020 The nonlinear optical susceptibility *Nonlinear Optics* (Amsterdam: Elsevier) pp 1–64
[29] Myers L E and Bosenberg W R 1997 Periodically poled lithium niobate and quasi-phase-matched optical parametric oscillators *IEEE J. Quantum Electron.* **33** 1663–72
[30] Powers P E, Kulp T J and Bisson S E 1998 Continuous tuning of a continuous-wave periodically poled lithium niobate optical parametric oscillator by use of a fan-out grating design *Opt. Lett.* **23** 159
[31] Smith D C 1977 High-power laser propagation: thermal blooming *Proc. IEEE* **65** 1679–714
[32] Chen B, Campos J F, Liang W, Wang Y and Xu C Q 2006 Wavelength and temperature dependence of photorefractive effect in quasi-phase-matched $LiNbO_3$ waveguides *Appl. Phys. Lett.* **89** 043510
[33] Buse K, Breer S, Peithmann K, Kapphan S, Gao M and Krätzig E 1997 Origin of thermal fixing in photorefractive lithium niobate crystals *Phys. Rev. B: Condens. Matter Mater. Phys.* **56** 1225–35
[34] Moretti L, Iodice M, Della Corte F G and Rendina I 2005 Temperature dependence of the thermo-optic coefficient of lithium niobate, from 300 to 515 K in the visible and infrared regions *J. Appl. Phys.* **98** 036101
[35] Wu X, Zhou C, Huang W R, Ahr F and Kärtner F X 2015 Temperature dependent refractive index and absorption coefficient of congruent lithium niobate crystals in the terahertz range *Opt. Express* **23** 29729

[36] Hobden M V and Warner J 1966 The temperature dependence of the refractive indices of pure lithium niobate *Phys. Lett.* **22** 243–4

[37] Verluise F, Laude V, Huignard J-P, Tournois P and Migus A 2000 Arbitrary dispersion control of ultrashort optical pulses with acoustic waves *J. Opt. Soc. Am.* B **17** 138–45

[38] Tournois P 1997 Acousto-optic programmable dispersive filter for adaptive compensation of group delay time dispersion in laser systems *Opt. Commun.* **140** 245–9

[39] Zelmon D E, Small D L and Jundt D 1997 Infrared corrected Sellmeier coefficients for congruently grown lithium niobate and 5 mol% magnesium oxide-doped lithium niobate *J. Opt. Soc. Am. B* **14** 3319

[40] Malitson I H 1963 A redetermination of some optical properties of calcium fluoride *Appl. Opt.* **2** 1103

[41] Malitson I H 1965 Interspecimen comparison of the refractive index of fused silica *J. Opt. Soc. Am.* **55** 1205

[42] Michael Bass E V S, DeCusatis C, Enoch J M, Lakshminarayanan V, Li G, MacDonald C and Mahajan V N 2009 *Handbook of Optics, Third Edition Volume IV: Optical Properties of Materials, Nonlinear Optics, Quantum Optics (set)* (New York: McGraw-Hill Professional)

[43] Wang Y *et al* 2020 1.1 J Yb:YAG picosecond laser at 1 kHz repetition rate *Opt. Lett.* **45** 6615–8

[44] Chi H, Wang Y, Davenport A, Menoni C S and Rocca J J 2020 Demonstration of a kilowatt average power, 1 J, green laser *Opt. Lett.* **45** 6803–6

Chapter 5

Advances of ultraviolet light sources: towards femtosecond pulses in the few-cycle regime

Vincent Wanie, Lorenzo Colaizzi, Andrea Cartella, Andrea Trabattoni and Francesca Calegari

5.1 Introduction

Ultraviolet (UV) radiation plays an important role in a plethora of physical, chemical, and biological processes occurring in nature [1]. In the last two centuries, scientists have introduced methods to artificially produce UV light to be used for multiple purposes, ranging from health-related technology [2, 3] to, more recently, 3D printing and lithography [4–6]. Since the advent of lasers and nonlinear optics [7], however, the generation of UV radiation has dramatically advanced with a large impact not only on technology and society but also on scientific research. Indeed, nowadays UV radiation is diffusely employed in laboratories to investigate photo–matter interactions and UV spectroscopy has been promoted as one of most compelling branches of modern science. As an example, fewer than 300 UV-related publications were released between 1920 and 1927 [8], but according to PUBMED, in the period between 1960 and 2001, almost 40 000 papers on the subject 'ultraviolet radiation' were published [8].

This chapter aims at an overview of only a small portion of such an immense landscape. In this context, we will focus on the generation of ultrashort UV laser pulses, i.e., with femtosecond (fs), few-cycle, temporal duration. Furthermore, we will describe the main applications of these pulses in the field of time-resolved UV spectroscopy.

In 1801, Johann Wilhelm Ritter conducted spectroscopical studies on the solar spectrum. He split sunlight with a prism and let the refracted radiation impinge onto a paper sample that was coated with silver chloride. It had been known since the 16th century that minerals containing silver chloride darkened upon light exposure. It also had been demonstrated that silver chloride irradiated by blue light became

doi:10.1088/978-0-7503-2536-3ch5

darker than when exposed to yellow light. Ritter observed that the strongest darkening chemical action on silver chloride occurred due to a portion of the solar spectrum that was lower in wavelength than the visible violet part.

Ritter dubbed such *invisible* rays 'de-oxidizing rays' to emphasize their chemical action on matter in contrast to the thermal effects induced by the 'heat rays', i.e., the radiation at the other end of the visible spectrum. Over time, the simpler term 'chemical rays' was adopted to describe the 'de-oxidizing rays' along with the adjacent violet-blue rays. The term 'chemical rays' remained popular throughout the 19th century, but it was eventually replaced by the more restrictive term ultraviolet (UV) radiation.

Generally, UV light is defined as electromagnetic radiation with a wavelength shorter than 400 nm. In nature, it is efficiently produced by the sun in the range between ~100 nm and 400 nm and has a strong impact on the chemistry of life. For example, the major source of vitamin D for most land vertebrates, including humans, comes from exposure to ultraviolet light at wavelengths between 290 nm and 315 nm [9], also known as UV-B radiation. In this context, the UV-B photons enter the skin and photolyze 7-dehydrocholesterol to previtamin D_3, which in turn is isomerized by the body's temperature to vitamin D_3. However, it is well known that UV radiation also represents the most harmful and mutagenic component of the solar spectrum [10]. Indeed, DNA efficiently absorbs UV-B and UV-A (wavelengths between 315 nm and 400 nm) photons, and the main photochemical reactions occur in the nucleobases, including thymine and cytosine [11]. The chemical effects of UV light on DNA are mostly explained in terms of the formation of dimeric photo-products (see figure 5.1). In this context, mutations in cultured cells and skin tumors were observed upon UV exposure [10, 12].

It is also interesting to note that many organisms have developed multiple photochemical schemes that employ visible light (especially blue photons) to repair UV-induced DNA damage. Such repairing methods are called photoreactivation [10]. The above-mentioned examples highlight the crucial role played by UV-induced mechanisms in photo-chemistry. The importance of ultraviolet light has been known for a long time and encouraged scientists to produce UV radiation artificially. For example, already at the end of the 19th century the sterilizing properties of short-wavelength light were known and artificial UV light became essential for health-related technologies.

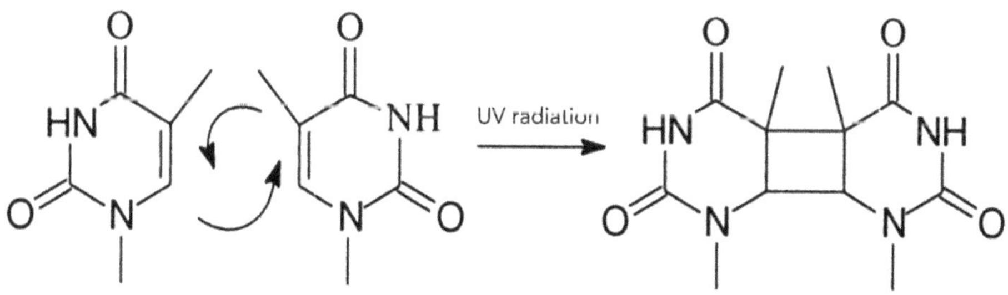

Figure 5.1. Formation of thymine dimers upon UV exposure. Reprinted from [12]. Copyright (2001), with permission from Elsevier.

In the last few decades, UV light has been promoted to be at the basis of important branches of human technology, such as 3D printing [4], photo-therapy [3], medical imaging [2], and lithography [5]. In this context the advent of lasers [13] has definitely played an important role to advance the generation of tailored UV radiation. Already in the 1960s, the first pulsed argon-ion lasers were able to generate UV wavelengths. Then, the development of excimer lasers established a routine technology to produce UV radiation. Argon-Fluoride (ArF) pulsed excimer lasers [14], for example, emit light at ~193 nm and are heavily used nowadays for integrated circuit production by photolithography [5]. Such pulses, although delivering several Joules per pulse, have a typical temporal duration of tens of nanoseconds and repetition rates limited to a few kHz. In order to obtain shorter pulse durations, second- or third-order nonlinear processes can be used to convert infrared or visible laser pulses into ultraviolet pulses. Using second- or third-harmonic generation (SHG and THG, respectively) tunable UV lasers with wavelengths down to $\simeq$ 200 nm and pulse durations down to a few femtoseconds have been obtained [15]. Few-fs, or even sub-fs, laser pulses at shorter wavelengths ($<$100 nm) are typically produced by strong-field driven high-harmonic generation (HHG) [16] or free-electron lasers (FELs) [17].

Nowadays, femtosecond UV pulses are mostly used in basic research, as unique tools to perform ultrafast spectroscopy of excited states of matter. In the field of femtochemistry, for example, they allow scientists to investigate in real time how specific chemical reactions build up in molecules and—potentially—control them with high temporal resolution. In this context, while tens of femtoseconds are sufficient to resolve the motion of atoms in a chemical reaction, few-femtosecond—or possibly sub-femtosecond—pulses are required to take snapshots of the electrons surrounding the nuclei and participating to the overall dynamics. Spectroscopy employing ultrashort UV pulses aims at investigating the photo-induced electron dynamics in excited targets, interrogating the interplay between the electronic and nuclear degrees of freedom, and—ultimately—tailoring the photo-interaction to control the functionality of the target itself.

In the next section, we will review the current existing methods for the generation of femtosecond UV pulses. Then, we will review the main applications of such pulses and the state-of-the-art of ultrafast UV spectroscopy.

5.2 Generation of femtosecond UV pulses

Nowadays, well established technologies based, for example, on Ti:Sapphire or ytterbium lasing materials allow for the production of commercial lasers delivering ultrashort near-infrared pulses, with a temporal duration of few tens of femtoseconds. However, such an achievement is extremely challenging at lower wavelengths. UV lasers, for instance, are limited to the use of narrowband lasing media, providing typical durations in the nanosecond timescale.

For the above-mentioned reason, ultrashort UV pulses are usually generated by femtosecond near-infrared pulses through nonlinear optical processes in different materials.

In this context, the up-conversion of the near-infrared photons into the UV spectral region results from the interaction between the driving laser and the polarization tensor of the nonlinear medium. In the perturbative regime, the polarization tensor $P(t)$ can be written as an electro-optical expansion as follows:

$$P(t) = \epsilon_0(\chi^{(1)}E(t) + \chi^{(2)}E^2(t) + \chi^{(3)}E^3(t) + \cdots), \tag{5.1}$$

where $E(t)$ is the driving electric field, and $\chi^{(n)}$ is the nth order electric susceptibility of the medium. If we assume that $E(t)$ oscillates at the central frequency ω, it is immediately evident from equation (5.1) that any nonlinear term, i.e., with $n > 1$, is responsible for the generation of new frequencies, which are possibly harmonics of ω. However, due to the perturbative nature of equation (5.1), the larger n is, the smaller the amplitude of $P^{(n)}$. In this context, the most direct—and intuitive—way of generating UV frequencies starting from a near-infrared laser is a third-order process, in which the $\chi^{(3)}$ coefficient of $P(t)$ plays a dominant role. Indeed, $E^3(t)$ intrinsically produces terms oscillating at 3ω, i.e., the third harmonic of ω, as schematically represented in figure 5.2(a). Such a process is usually dubbed direct THG and it is relevant in centrosymmetric media in which $\chi^{(2)} = 0$. Through THG, for example, light at $\lambda = 800$ nm can be up-converted into radiation at $\lambda = 266$ nm.

The role of P as a possible source of electromagnetic radiation becomes evident when equation (5.1) is combined with Maxwell's equations. In the particular case of a $\chi^{(3)}$-driven process, the following wave equation can be written:

$$\nabla^2 E - \frac{\epsilon_r}{c^2}\frac{\partial^2 E}{\partial t^2} = \frac{\partial^2 P^{(3)}}{\partial t^2}, \tag{5.2}$$

where c is the speed of light, ϵ_r is the medium's relative permittivity and $P^{(3)} = \epsilon_0 \chi^{(3)} E^3$ is the leading nonlinear order of the polarization tensor. Equation (5.2) shows that the nonlinear polarization acts as a source of electric field. In this framework, THG can be interpreted in terms of energy conservation as the conversion of three waves at frequency $\omega_1 = \omega_2 = \omega_3 = \omega$ into a fourth wave of frequency $\omega_4 = 3\omega$, as reported in figure 5.2(b). In order to have a better understanding of the THG process, it's beneficial to develop equation (5.2) in the

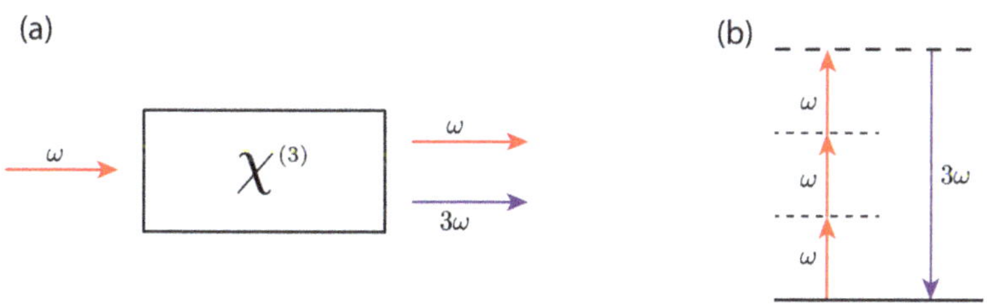

Figure 5.2. Principle of parametric third-harmonic generation: (a) Propagation of waves with frequency ω into a nonlinear medium allows the generation of waves with frequency 3ω through the third-order susceptibility $\chi^{(3)}$. (b) Energy conservation: Three waves with the same frequency ($\omega_1 = \omega_2 = \omega_3 = \omega$) are converted into a frequency tripled wave $\omega_4 = 3\omega$.

frequency domain. If we assume the electric field propagating along the axis z, and ω_0 to be the carrier frequency, the spectrum of E can be written as:

$$E(z, \omega) = A(z, \omega - \omega_0)e^{ikz}, \tag{5.3}$$

where A is the spectral amplitude. Consequently, equation (5.2) can be rewritten as:

$$\frac{\partial^2 A}{\partial z^2} + 2ik\frac{\partial A}{\partial Z} + [k^2(\omega) - k^2]A = \omega^2 P^{(3)}(z, \omega)e^{-ikz}, \tag{5.4}$$

where

$$k^2(\omega) = \epsilon_r(\omega)\frac{\omega^2}{c^2} \tag{5.5}$$

accounts for the frequency-dependent dispersion in the medium. It is convenient to expand $k(\omega)$ in a power series as follows:

$$k(\omega) = k_0 + k_1(\omega - \omega_0) + \frac{1}{2}k_2(\omega - \omega_0)^2, \tag{5.6}$$

where k_1 and k_2 represent the inverse of the group velocity ($k_1 = \partial k/\partial\omega = 1/v_g$) and the group velocity dispersion (GVD) of the carrier frequency ($k_2 = \partial^2 k/\partial\omega^2$), respectively. Since these two coefficients set the dispersion of the generated third harmonic, they play a crucial role in the choice of the THG nonlinear medium, as will be discussed in the next sections. Combining equations (5.6) and (5.4) and Fourier transforming into the time domain, the wave equation for the electric field reads as:

$$\left[\frac{\partial^2}{\partial z^2} + 2ik_0\left(\frac{\partial}{\partial z} + \frac{1}{v_g}\frac{\partial}{\partial t}\right) - k_0 k_2\frac{\partial^2}{\partial t^2}\right]Ae^{i(kz-\omega t)} = \epsilon_0\chi^3\frac{\partial}{\partial t^2}A^3 e^{i(kz-\omega t)}. \tag{5.7}$$

Equation (5.7) can be simplified by employing the slowly varying envelope approximation (SVEA), which is valid when the variation of the temporal envelope $A(z, t)$ is slower than the carrier frequency, i.e., when $|\partial^2 A/\partial z^2| \ll |k_0\partial A/\partial z|$. In this framework, the first term of equation (5.7) is neglected. However, SVEA is no longer applicable for few-cycle pulses, in which the duration of $A(z, t)$ becomes comparable with the optical cycle of the carrier wave. For such a case, Brabec and Krausz introduced the so-called slowly evolving wave approximation (SEWA) [18], that holds if:

$$\left|\frac{\partial E}{\partial z}\right| \ll |k_0 E|. \tag{5.8}$$

It requires that both $A(z, t)$ and the phase of the carrier do not vary significantly along a distance $d_0 = \lambda_0 = 1/k_0$. Although being more stringent than SVEA, SEWA holds for ultrashort laser pulses and allows equation (5.7) to be rewritten as:

$$\left[\left(\frac{\partial}{\partial z} + \frac{1}{v_g(3\omega)}\frac{\partial}{\partial t}\right) + \frac{ik_2(3\omega)}{2}\frac{\partial^2}{\partial t^2}\right]A_{3\omega} \propto \chi^3 A_\omega^3 e^{i\Delta kz}. \tag{5.9}$$

As mentioned above, the left-hand side of equation (5.9) describes the propagation of the generated field at 3ω, with the first- and second-order dispersion (via $v_g(3\omega)$ and $k_2(3\omega)$, respectively). The right-hand side instead represents the source of the third-harmonic field and depends on a range of variables affecting the THG efficiency, namely the third-order susceptibility of the generating medium χ^3, the amplitude A_ω of the driving pulse, and the phase mismatch $\Delta k = k_{3\omega} - 3k_\omega$.

Among them, Δk is key to the overall efficiency of the χ^3-driven process. Let's assume the propagation in the nonlinear medium to be of finite length L and the dispersion term in equation (5.9) to be negligible. Under these assumptions, the THG efficiency, i.e., the ratio between the intensity of the field at 3ω $I(3\omega, L)$, and intensity of the driving field, $I(\omega, L)$, can be written as:

$$\eta_{\mathrm{THG}} = \frac{I(3\omega, L)}{I(\omega)} \propto |\chi^{(3)}|^2 L^2 I^2(\omega)\mathrm{sinc}^2\left(\frac{\Delta kL}{2}\right). \tag{5.10}$$

In order to maximize the THG efficiency, the term $|\chi^{(3)}|^2 L^2 I^2(\omega)$ has to be maximized and the phase-matching condition $\Delta k = 0$ needs to be satisfied.

In the next paragraphs, several THG methods that aim at achieving such a result will be presented, and the advantages and disadvantages of each method will be discussed.

5.2.1 UV generation in crystals

In nonlinear crystals, the phase mismatch term Δk (equation (5.10)) can be minimized through their birefringent property: crystal axes with distinct refractive indexes are exploited to configure the relative polarization orientation between each of the frequency components. This allows for matching the phase velocity of the fundamental and THG fields over a long propagation distance where the signal builds up, thereby significantly increasing the THG efficiency.

While direct THG can be phase matched in this manner, χ^3 coefficients in centrosymmetric crystals are relatively small compared to χ^2 coefficients in non-centrosymmetric ones, limiting the conversion process. For this reason, typical THG setups are based on cascaded second-order nonlinear processes exploiting the larger χ^2 coefficients: SHG ($\omega + \omega = 2\omega$) is followed by a sum frequency generation (SFG) with the fundamental field ($\omega + 2\omega = 3\omega$). The conversion efficiency using this scheme is typically almost one order of magnitude higher than for direct THG ($\omega + \omega + \omega = 3\omega$) [19].

The schemes described above are commonly used for generating UV pulses with duration ranging from few tens of femtoseconds to several picoseconds. However, two important factors must be further considered for the generation of few-femtosecond UV pulses. First, the GVD of crystals in the UV spectral range is high. This inevitably induces a strong positive chirp over propagation in the generating medium. The lack of materials providing negative GVD in the UV forces the need for very complex dispersion compensation techniques, which are usually based on prisms and grating elements and therefore add a significant level of difficulty and complexity to the UV generation setups [20]. Second, the dispersion properties of crystals limit the range of wavelengths that can be efficiently phase matched due to the refractive index dependence of the relation Δk. The phase-matching bandwidth—i.e., the maximum bandwidth that can be phase matched in the THG process—is expressed by

$$\Delta\lambda \propto \frac{1}{\left|\frac{1}{v_{3\omega}} - \frac{1}{v_{\omega}}\right| L} \tag{5.11}$$

and it is inversely proportional to both the crystal length and the group velocity mismatch in the medium, associated to the group velocity of the fundamental ν_{ω} and third-harmonic $\nu_{3\omega}$ components. Equation (5.11) implies that in order to generate the large bandwidth required for few-cycle UV pulses it is necessary to drastically reduce the crystal thickness, thereby reducing the conversion efficiency (equation 5.10). This effect is intrinsic to all nonlinear processes and for this reason crystals are not ideal candidates for the generation of few-femtosecond UV pulses.

However, μJ level UV pulses with duration below 20 fs can be obtained by frequency doubling ultrashort visible pulses [20, 21] generated by a non-collinear optical parametric amplifier (NOPA). In a NOPA, a small portion of the fundamental frequency ω (usually around 800 nm) is spectrally broadened down to the visible part of the electromagnetic spectrum by self-phase modulation (a χ^3 process) and subsequently amplified in beta barium borate ($\beta - BaB_2O_4$, usually referred to as BBO) nonlinear crystals. Because of the dispersion properties of BBO crystals in the infrared and visible spectral regions and because of the non-collinear geometry that allows for better exploiting the crystal birefringent properties, a very broad bandwidth can be amplified in crystals as thick as a few millimeters, yielding a high 800-nm-to-visible conversion efficiency. Nevertheless, in order to generate a 5 fs UV pulse by frequency doubling a visible one, a crystal thickness of only 5 μm would be required, yielding a minimal conversion efficiency [21]. In order to partially mitigate this limitation, advanced techniques ultimately based on the control of the bandwidth and spectral phase of the generating pulses have been demonstrated.

Baum and coworkers [22] achieved a record pulse duration of 7.1 fs for UV pulses generated in crystals due to achromatic phase matching. In this technique, the output of a visible NOPA is traveling through a pair of prisms designed to obtain a collimated beam of spatially dispersed wavelengths. When such beam is focused on a BBO to frequency double it, different wavelengths impinge on the crystal at different angles, and this enables satisfying the phase-matching condition for a much larger bandwidth. In fact, the phase-matching bandwidth is increased by a factor of 80 with respect to a typical collinear focusing geometry. For this reason, a considerably thicker BBO crystal can be employed, yielding a sizable conversion efficiency of 20% and producing 250 nJ pulses with the aforementioned 7.1 fs duration.

More recently, Borrego Varillas and coworkers [23] proposed another approach, also based on the frequency doubling of a NOPA in the visible, achieving the generation of sub-10 fs UV pulses with 1.5 μJ pulse energy (i.e., with around 10% efficiency). In this case, the output of the NOPA has been frequency up-converted by SFG with a narrowband pulse at the fundamental frequency of 800 nm. Because of the indirect phase transfer mechanism [24], the SFG process transferred the NOPA's nearly flat spectral phase to the UV pulses, which were almost transform-limited. Therefore, even though the amplified bandwidth was smaller than that obtained by Baum and coworkers, a similar pulse duration could be achieved with a simpler and more robust setup.

5.2.2 UV generation in gas cells

To circumvent the limitations imposed by crystals for the generation of few-cycle UV pulses, alternative nonlinear media must be considered. Table 5.1 reports the ionization potential, GVD at 266 nm and third-order susceptibility (χ^3) of the noble gases helium, neon, argon, krypton and xenon. For comparison, common optical materials such as BBO and fused silica are included. With GVD values that are at least three orders of magnitude lower with respect to BBO and fused silica, the quasi dispersion-free property of gases are of potential interest for the generation of broadband and compressed UV pulses. As an example, the dispersion introduced by a 100 μm thick BBO crystal to a transform-limited 3 fs pulse stretches its duration to roughly 50 fs. In contrast, the pulse duration remains unaffected when such short pulses travel through noble gases: 1 mm of Xe—the gas with the highest GVD among those reported in table 5.1—would increase the duration of 3 fs pulses by less than 0.02%.

Nevertheless, the low χ^3 values of noble gases reported in table 5.1 relative to the optical materials show that low dispersion comes at a cost of low conversion efficiency. The latter must be compensated by increasing the density of emitters in

Table 5.1. List of nonlinear medium and their relevant properties for the generation of few-femtosecond UV pulses via direct THG. GVD values are taken from [25]. Third-order susceptibilities for He, Ne, Ar, Kr, Xe, and beta barium borate (BBO) are given for a 1055 nm driving laser [26, 27]. $\chi^{(3)}$ for fused silica is given at 1065 nm [28].

Nonlinear medium	Ionization potential (eV)	GVD at 266 nm (fs^2 mm^{-1})	$\chi^{(3)}$ (m^2 V^{-2})
He	24.59	0.003 6607	3.43×10^{-28}
Ne	21.56	0.009 1518	6.17×10^{-28}
Ar	15.76	0.082 890	8.06×10^{-27}
Kr	14.00	0.192 19	2.20×10^{-26}
Xe	12.13	0.591 99	6.46×10^{-26}
Fused silica		197.53	2.0×10^{-22}
Beta barium borate (BBO)		524.76	4.4×10^{-23}

the generating medium in order to obtain sufficient UV pulse energies for experiments.

Seminal work for generating few-femtosecond UV pulses has been performed by Graf *et al* in 2008 [29]. Through direct THG in gas, 3.7 fs UV pulses were obtained, spanning a spectral range between 230–310 nm, as shown in figure 5.3(b). The generation was driven by 6 fs near-infrared pulses focused in a 2.5 mm-diameter quasi-static gas cell filled with 5 bar of neon. The suppression of the fundamental driving field was achieved by using a silicon wafer at Brewster's angle, reducing the remaining infrared component by more than two orders of magnitude, at the expense of collecting only 60% of the UV radiation generated at the source.

This result was achieved after a careful investigation of different noble gases, as shown in figure 5.3(a). The UV pulse energy is reported as a function of the gas pressure. For xenon and argon, the pulse energy shows a clear maximum before degrading when an optimal gas pressure has been reached. The effect was attributed to ionization of the medium, limiting the achievable pulse energy that is competing with plasma generation. Instead, because of the higher ionization potential, a pulse energy of ~1.4 μJ (conversion efficiency of 0.6%) was obtained in neon. However, the lower $\chi^{(3)}$ has to be compensated with a higher pressure, which is about five times higher than for argon. For the same reason, an optimal gas pressure was not observed in helium since the high gas load was not compatible with the vacuum system of the UV generation setup.

While ionization can affect the UV generation by drastically reducing the achievable pulse energy, it can also be exploited to shape the driving field and obtain a larger bandwidth than the one expected from a pure THG process. Indeed, compared to the duration of the driving pulse, the duration of UV pulses generated via direct THG is expected to be shorter by a factor $\sqrt{3}$, as can be deduced by the presence of $E^3(t)$ in equation (5.1). Therefore, 5 fs driving pulses are expected to generate UV pulses of duration $\tau_{3\omega} = \tau/\sqrt{3} = 3$ fs, as demonstrated by Reiter *et al* [30]. In addition, they also investigated the effect of additional nonlinear processes contributing to further spectral broadening by developing a nonlinear propagation

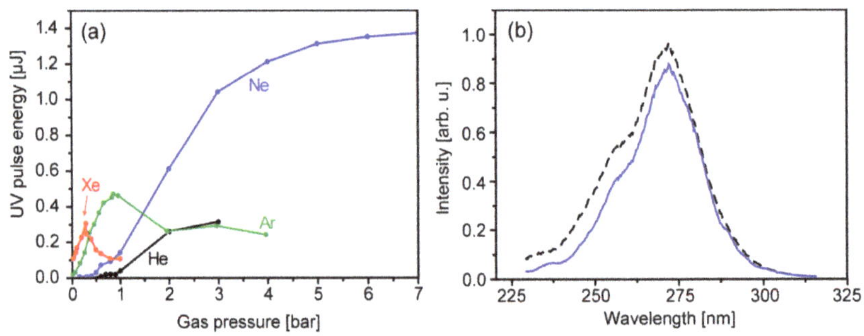

Figure 5.3. (a) Energy of UV pulses generated by focusing a 6 fs, 780-nm pulses into a quasi-static gas cell versus gas pressure for xenon, argon, neon, and helium. (b) Deep ultraviolet spectra obtained with 5 bar of neon supporting a transform-limit duration of 3.4 fs measured directly at the source (black curve) and after reflections on the Si wafer and aluminum mirrors (blue). Reprinted with permission from [29] © 2008 The Optical Society.

model for the driving and the generated pulses, including notably the Kerr effect, self-steepening and ionization through Keldysh-type equations. The critical role of ionization is highlighted in figure 5.4. The spatial profile for both the driving field and the generated UV pulses is simulated as a function of time. The first row ((a) and (b)) shows the results when ionization of the gas medium by the driving field is not taken into account, while the bottom row ((d) and (e)) shows the results in the presence of ionization. Comparing each row, it is observed that ionization induces reshaping of the spatio-temporal profiles: in (d), the intensity of the driving field is confined to the leading edge, while it decreases rapidly on the trailing edge. As the increasing intensity ionizes the medium, a steep plasma profile is created. By acting as a negative lens for the trailing edge, an overall defocusing effect is obtained. In turn, the spectral broadening of the UV pulse is confined only to the leading edge, resulting in a shorter pulse (f) compared to the ionization-free case (c), entering the sub-fs time domain. Ionization in this case is therefore working as a temporal gate for the generation and, in theory, even allows the generation of attosecond pulses in the UV spectral range.

In 2019, one of the broadest continuums in the ultraviolet has been obtained experimentally, with a bandwidth ranging from 210–340 nm and supporting sub-2-fs pulses (figure 5.5) [15]. In this case, a near 5 fs near-infrared pulse centered at 770 nm was focused in a laser-machined fused silica gas cell filled with 1.1 bar of argon. An important technological aspect in this work is the fabrication method of the glass cell, known as FLICE (femtosecond laser irradiated and chemical etching) [31]. The technique is commonly employed for the fabrication of waveguides and photonic integrated circuits and allows one to manufacture complex shapes with a resolution

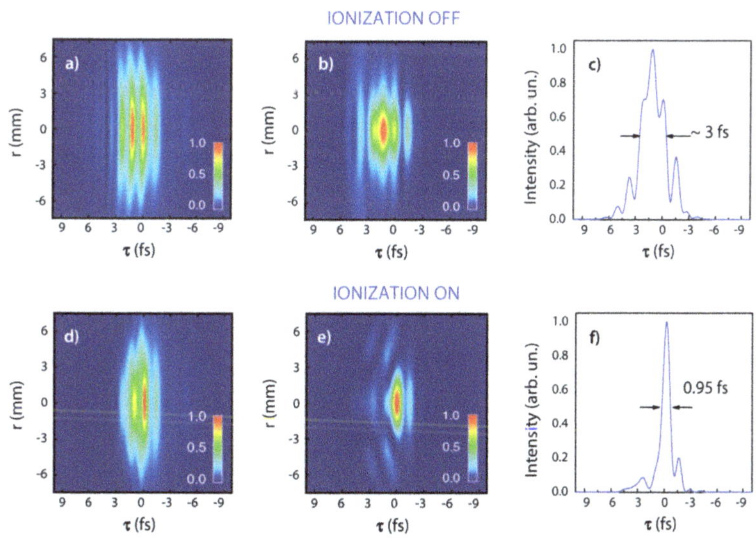

Figure 5.4. Simulated far-field spatio-temporal profile of the driving field (a, d) and the generated THG field (b, e) in a gas cell filled with 9 bar of neon, with and without including ionization-induced self-defocusing of the driving field. (c, f) Corresponding integrated UV temporal profile, having UV pulse duration (FWHM) of 3 fs (c) and 0.95 fs (f). Adapted with permission from [30] © 2010 by the American Physical Society.

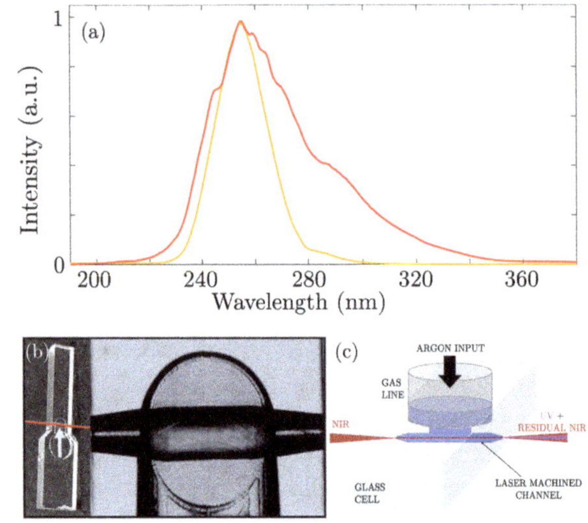

Figure 5.5. (a) THG in the ultraviolet obtained by focusing 5 fs short, 770 nm laser pulses in a laser-micromachined gas cell filled with 0.2 bar (yellow) and 1.1 bar (red) argon gas. The latter spectrum supports a transform-limit duration of 1.45 fs. (b) Left: Picture of the glass cell used. The red line represents the laser propagation direction while the white arrow represents the gas flow direction. Right: A magnified image of the channel acquired with a microscope. The extremities of the channel are reduced in a conical shape with a 400 μm diameter in order to reduce the out-flowing gas. (c) Schematic representation of the gas cell showing the laser beam, the gas inlet, and the channel shape. Adapted with permission from [15] © 2019 The Optical Society.

of a few μm [32]. In the context of nonlinear interactions with a dense gas medium, adjusting the design of the cell allows for a fine control over the gas distribution and/or the guiding properties in order to optimize the nonlinear processes at play. Figure 5.5(b) shows an image of the cell employed in [15], which consists of a 3 mm long channel, connected to the gas supply line. With a maximum diameter of 1 mm at the center, the extremities of the channel are decreased down to 400 μm, allowing reaching several bars of gas within the cell while limiting the out-flowing gas in the surrounding high-vacuum environment of the UV generation setup. The UV light source providing few-femtosecond pulses is reported to be synchronized with attosecond XUV pulses, promising a temporal resolution that can be exploited for several applications in ultrafast molecular science. An overview of these applications is given in the last section of this chapter.

5.2.3 UV generation in fibers and capillaries

A different approach for the generation of ultrashort UV pulses is based on capillaries, also referred to as hollow-core fibers (HCF). Capillaries allow the guided propagation of laser pulses over long distances with an energy that is spatially confined in order to reach a high intensity. Filled with noble gas, the interaction length for nonlinear processes is greatly increased, partially mitigating one of the main drawbacks of the generation in gas cells described above. The wavevector for light propagating in gas-filled capillaries is described by

$$k(\lambda) \approx \frac{2\pi}{\lambda} + \frac{2\pi P\delta(\lambda)}{\lambda} - \frac{u_{\mathrm{nm}}^2\lambda}{4\pi a^2}, \tag{5.12}$$

where P is the gas pressure in the capillary, δ is related to the refractive index (n) of the gas ($\delta = (n - 1)/P$), u_{nm} is a constant depending on the spatial mode that is propagating in the capillary and a is the diameter of the capillary itself. The three terms in equation (5.12) represent the contributions to the wavevector due to the free space propagation, the gas dispersion, and the waveguide dispersion, respectively. When the capillary is used to generate UV by THG, equation (5.12) holds for both the fundamental field at frequency ω and third-harmonic field at frequency 3ω, and the choice of gas species, gas pressure and waveguide diameter can be used to achieve the phase-matching condition [33].

Pioneering experiments of UV generation in capillaries were performed roughly two decades ago, showing the possibility of generating sub-10 fs pulses either by cross-phase modulation (a χ^3 process) [34], or by using cascaded nonlinearities [35]. More recently, the effect of the waveguide on the dispersion, together with the nonlinear propagation effects in the gas, have been fully exploited to achieve the propagation of solitons (i.e., waves propagating without losses and dispersion) in the capillaries [36]. This has been shown to generate self-compressed pulses in the near-infrared and visible spectral region, while simultaneously producing a broadband spectrum in the UV [37], capable of supporting few-fs pulses. Research on this emerging technology is very active, and additional developments based, for example, on using a pressure gradient in the capillaries have been demonstrated [38].

One particularly interesting aspect of UV generation in HCFs is the recently demonstrated possibility of generating circularly polarized pulses. Because the phase matching in HCFs relies on the gas pressure and capillary diameter rather than on birefringence (such as in crystals), the driving pulses are not required to be linearly polarized. For instance, four-wave mixing (FWM, a χ^3 process) using circularly polarized pulses at 800 nm and 400 nm have been used to generate ultrabroadband circularly polarized UV spectra, supporting pulse durations down to 4 fs [39]. Such control over the polarization state of ultrabroadband UV pulses opens the way to several experiments in ultrafast molecular science that cannot be realized with linearly polarized pulses, as will be detailed later in this chapter.

A different method for the generation of few-femtosecond UV pulses is based on hollow-core photonic crystal fibers (HC-PCFs). These are capillaries where the hollow core is surrounded by a microscopic array of holes whose structure can be chosen to achieve specific guiding and dispersive properties. On one hand, because of the additional degree of freedom represented by their design, HC-PCFs can be tailored to achieve a high conversion efficiency and broad bandwidth in the UV generation even for a relatively small input pulse energy (on the order of the μJ) and relatively long duration (on the order of a few tens of femtoseconds) [40, 41]. On the other hand, because of their microscopic structure, HC-PCFs are intrinsically more delicate than HCFs, and their usage cannot be scaled up to very high pulse energies. Nevertheless, they represent a powerful tool to scale up the repetition rate of experiments by employing oscillators or fiber-based laser systems.

5.2.4 Characterization of few-fs UV pulses

For ultrabroadband UV pulses, common characterization techniques such as SHG-FROG (frequency-resolved optical gating) [42] or SPIDER (spectral phase interferometry for direct electric-field reconstruction) [43] cannot be applied considering the absence of nonlinear medium in the UV for SHG. In addition, due to the strong second-order dispersion of most materials in the UV, dispersion-free techniques need to be employed to preserve the short pulse duration in the characterization setup. Furthermore, the very low efficiency of nonlinear processes discussed in the previous sections requires specific characterization methods which account for relatively low photon flux. It is no surprise that the works reporting few-femtosecond UV pulses have employed various alternative characterization methods.

Complete amplitude and phase reconstruction has been performed, for example in [23], using a two-dimensional spectral shearing interferometry (2DSI) technique [44] and in [29], adapting a self diffraction FROG [45] scheme in vacuum to overcome the dispersion limitations. More recently, Brahms *et al* in [36] employed an in-vacuum cross-correlation frequency-resolved optical gating (XFROG) specifically designed for the characterization of few-fs UV pulses generated in a HCF. Briefly, the generated pulse is combined with a narrowband gate pulse at 800 nm—obtained by spectrally filtering the fundamental field—in a BBO crystal optimized for difference frequency generation (DFG) as nonlinear interaction. Choosing DFG instead of SFG has two main advantages: firstly, the XFROG signal is generated in a spectral range around 400 nm in which BBO is transparent. Secondly, for a crystal of 5 μm thickness, sufficient phase-matching bandwidth is achieved by tuning the crystal angle in order to cover the full UV spectrum. Finally, employing advanced retrieval algorithms, such as ptychographic techniques commonly used in FROG analysis [46], allows one to measure the amplitude and the spectral phase of 3 fs UV pulses.

Alternatively, ionization-based autocorrelation [47] or cross-correlation schemes [15] have been performed using noble gas as a nonlinear medium. In both methods, the photoionization ions yield of a gas target is measured as a function of the delay between the generated UV pulse and its replica (autocorrelation) or the fundamental (cross-correlation). It is noteworthy that these methods do not allow for a full reconstruction of the field, instead they are based on assuming a temporal profile for the pulses with no information on the phase to be extracted.

For instance, in [15], xenon photoions' yield has been measured with a time-of-flight (TOF) mass spectrometer as a function of the UV and the fundamental near-infrared (NIR) delay, as represented in figure 5.6. In this case, for negative delays the UV pulse arrives before the NIR pulse. Here, the cross-correlation signal exhibits an overall step-like dynamics that was analyzed considering the electronic energy structure of xenon. Xenon photoexcitation cross-section and NIR intensity scaling allowed narrowing down the ionization channels to only two contributions: direct

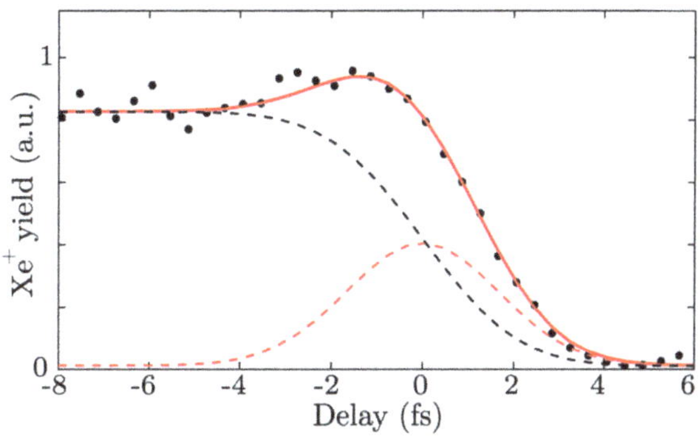

Figure 5.6. Measured xenon ions' yield as a function of the UV-NIR delay (black dots). For negative delays, the UV pulse precedes the NIR pulse. The dashed lines represent two distinct ionization channels which contribute to the delay dependent signal: the red dashed line shows simultaneous absorption of 2 UV + 2 IR photons, while the black dashed line corresponds to the resonance-enhanced transition. The red solid line represents the sum of these two contributions. Adapted with permission from [15] © 2019 The Optical Society.

2 UV + 2 NIR photons ionization versus a resonance-enhanced excitation (with nanosecond lifetime) followed by photoionization. These two ionization channels are respectively represented by the red and the black dashed lines in figure 5.6: the red dashed line represents the simultaneous UV and NIR absorption at the time overlap and was modeled as a Gaussian. The black dashed line represents the absorption through a Rydberg series which is accessible by two UV photons' absorption, followed by NIR photoionization, and is described by an error function. Therefore, the two contributions have been modeled with the following function:

$$f(t) = A_1 \exp\left(-\frac{(t-t_0)^2}{2\sigma^2}\right) + A_2 \operatorname{erf}\left(\frac{(t-t_0)}{\sqrt{2}\sigma}\right), \tag{5.13}$$

in which t_0 represents the time offset corresponding to the temporal overlap between the UV and NIR pulses, while the parameter σ represents the cross-correlation width. The UV width σ_{UV} was retrieved according to the relation:

$$\sigma = \sqrt{\frac{\sigma_{\mathrm{UV}}^2}{n_{\mathrm{UV}}} + \frac{\sigma_{\mathrm{NIR}}^2}{n_{\mathrm{NIR}}}}, \tag{5.14}$$

in which σ_{NIR} is the NIR width that can be measured independently, n_{UV} and n_{NIR} represent the number of UV and NIR photons involved for a specific ionization channel. The above-mentioned work characterized a UV pulse duration of 1.9 fs and provides a clear example that although dispersion-free, such method relies on the careful identification of relevant nonlinear order processes leading to photoionization.

Finally, it is worth mentioning a promising alternative gas-based approach recently reported by Kida, demonstrating transient grating XFROG using a thin noble gas target in vacuum (instead of air) [48]. This scheme combines the full reconstruction of amplitude and phase of an XFROG scheme with a dispersion-free medium for the nonlinear interaction.

5.3 Applications in ultrafast molecular science

Few-cycle UV pulses are unique tools to reproduce photochemical and photobiological processes in laboratories that play a key role in nature. Such studies require a few-femtosecond or even attosecond time resolution. Indeed, the above-mentioned timescales are essential to follow in real time not only the nuclear rearrangements in molecules following the photo-interaction, but also the even faster out-of-equilibrium electron motion. Only by probing these entangled dynamics it is possible to have a deep understanding of the photo-induced response of the target, and to explore possible ways of photochemical control.

Nowadays, time-resolved spectroscopy routinely employs light pulses in a vast range of the electromagnetic spectrum, with wavelengths covering the terahertz to the hard x-ray spectral regions. However, ultrafast spectroscopical studies employing few-cycle UV pulses are still extremely sparse. On one hand, this is a direct consequence of the many current technological challenges for the generation and manipulation of ultrabroadband few-cycle UV pulses, as discussed in the previous section of this chapter. On the other hand, ultrafast UV spectroscopy is key to molecular science: in several organic and bio-relevant molecules, mostly composed of carbon and nitrogen atoms, there are chromophores made of double and/or triple chemical bonds that can be easily photo-excited upon UV irradiation. Such a photo-response can be observed, for example, in the nucleic acid bases cytosine, adenine, thymine, guanine, and uracil with strong absorption bands between 230 and 290 nm [11], and in amino-acids such as tryptophan (Trp) and phenylalanine (Phe) [49, 50]. Several metallic complexes, of potential interest for the development of light-harvesting devices, also show absorption bands in the ultraviolet spectral region.

In the following sections, we provide an overview of the state-of-the-art of ultrafast UV spectroscopy on molecular systems, with particular emphasis on the properties of ultrashort UV pulses employed in the experiments: the time resolution, the bandwidth, and the polarization state. Perspectives for upcoming development will also be discussed.

5.3.1 Time resolution

The common approach for time-resolved measurements in molecular science utilizes the pump-probe scheme: a first light pulse (pump) excites a molecule, triggering an ultrafast dynamical process. A delayed pulse (probe) then captures a snapshot of the time-evolving system at a given instant. A series of snapshots taken at different pump-probe time delays provide information on the dynamical properties of the molecule. By combining near-infrared pump pulses with attosecond XUV or soft x-ray pulses, experiments provide the way to investigate nuclear rearrangements and electron dynamics in matter with few-femtosecond to attosecond time resolution [51]. Typically, the time resolution is given by the convolution between the pump and the probe pulses. Therefore, reducing the wavelength of few-cycle pulses from the near-infrared (1 cycle @ 800 nm = 2.67 fs) towards the ultraviolet (1 cycle @ 266 nm = 0.89 fs) intrinsically enhances the time resolution of an experiment.

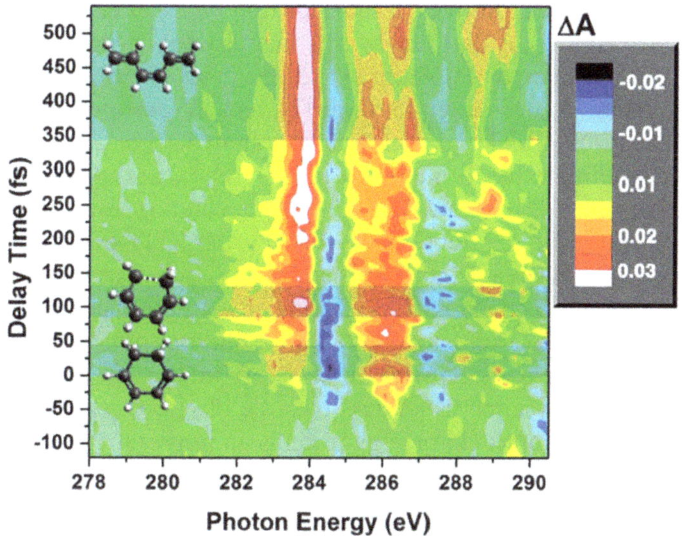

Figure 5.7. A false-color surface map showing the change in the x-ray absorption (ΔA) as a function of photon energy and delay time, with positive delay times indicating that the soft x-ray pulse follows the UV pulse. An increase in absorption is represented in red or white, a decrease in blue or black. Three temporal windows are indicated by the shaded regions, which each represent an important step in the ring-opening reaction process. Adapted from [52]. Reprinted with permission from AAAS.

The challenge of exploiting a pump-probe configuration combining femtosecond ultraviolet pulses with an attosecond light source based on HHG has been overcome very recently [52–54]. One of the first proof of principle experiments has been to track in real time the UV-induced ring-opening reaction of 1,3-cyclohexadiene, the prototype of an important category of reactions in organic chemistry. Indeed the study of this process served for supporting the theory of molecular orbitals through the elaboration of the Woodward–Hoffman rules that dictate the outcome of pericyclic reactions [55, 56], which led to the 1981 Nobel Prize in chemistry. Figure 5.7 shows the results of the seminal transient absorption experiment by Attar *et al* [52]. Photoexcitation by a 266 nm UV-pump pulse initiates the ring-opening reaction of 1,3-cyclohexadiene. To follow the nuclear rearrangement, the absorption of delayed soft x-ray pulses (ΔA) in the vicinity of the carbon *K* edge (284 eV) serves as a probe. The three shaded areas were identified to contain the signature of the corresponding molecular geometries shown on the left side of the figure.

Occurring over few hundreds of femtoseconds, the ring opening of 1,3-cyclohexadiene was observed by Attar *et al* using ~110 fs UV-pump pulses (temporal resolution of ~120 fs). However, a large class of molecular dynamics takes place on a much faster timescale, notably electron dynamics and their interaction with nuclear degrees of freedom near conical intersections. In 2019, Latka *et al* reported a significant improvement in terms of temporal resolution by employing ~10 fs UV pulses with XUV attosecond pulses to study the photodissociation dynamics of ozone molecules [53]. The photolysis of ozone by solar ultraviolet radiation is one of the direct mechanisms responsible for the depletion of the atmospheric ozone layer. A sketch of the experiment that reproduces this process is shown in figure 5.8(a). A ~10 fs UV pulse centered at 268 nm excites the ground state (*X*) of the molecule to a weakly binding excited state (*B*). Over few tens of femtoseconds, a fraction of the nuclear wavepacket undergoes molecular fragmentation along the vibrational coordinate R_1, leading to the formation of oxygen (O_2) that is probed by attosecond XUV pulses. Figure 5.8(b) shows the temporal evolution for the appearance of photoelectrons emitted from the O_2 molecules by the XUV-probe pulses, showing that the photolysis takes place on a timescale of only few tens of femtoseconds.

The two works described above illustrate the need to produce shorter and shorter UV pulses in order to have sufficient temporal resolution to disclose the mechanism of photo-induced processes occurring at ultrafast timescales. In this perspective, attosecond spectroscopy aims to address the role of electron dynamics in matter. Providing only one example, charge migration is an important phenomenon in bio-relevant molecules that is purely driven by electrons: a charge density is rapidly

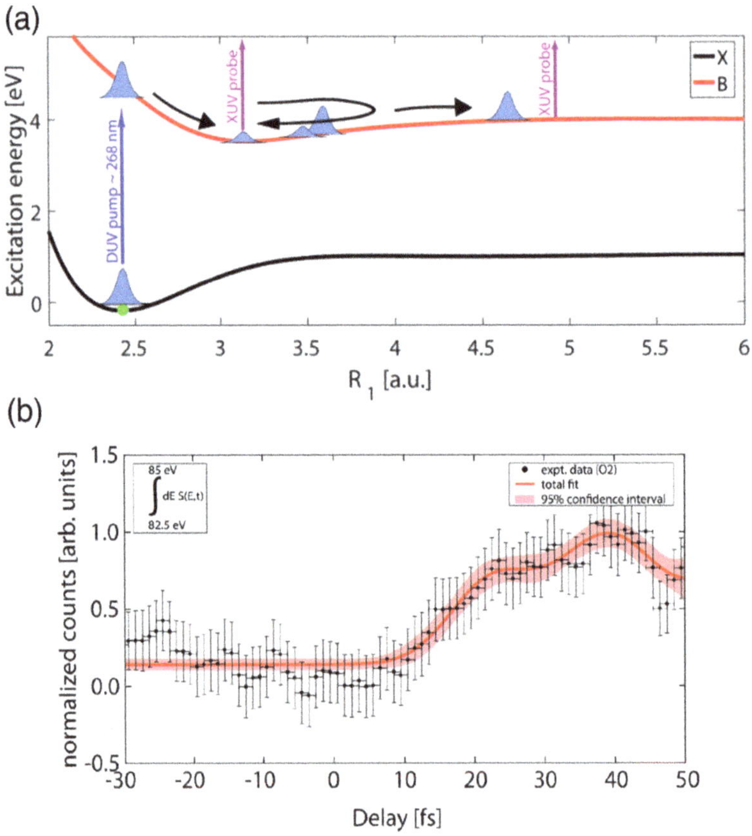

Figure 5.8. (a) Sketch of the UV-induced photodissociation dynamics of ozone probed by XUV attosecond pulses. (b) Photoelectron yield from oxygen (O_2) molecules, produced during the photolysis of ozone. Adapted with permission from [53]. © 2019 by the American Physical Society.

redistributed, or oscillates back and forth along a single molecule or within a molecular chain in less than 5 fs [57]. This charge migration process was notably identified experimentally in photoionized molecules [58, 59]. The use of ultrashort UV-pump pulses for photoexcitation may provide the opportunity to observe charge migration in neutral molecules, which is a step closer towards reproducing and elucidating photochemical and photobiological processes.

5.3.2 Bandwidth

Advanced light sources delivering few-femtosecond UV pulses also provide access to an intrinsic ultrabroadband spectrum via the time-bandwidth product:

$$\Delta\tau \times \Delta\lambda \times \frac{c}{\lambda^2} \approx 0.44 \text{ (Gaussian)}. \tag{5.15}$$

For instance, with a central wavelength $\lambda = 266$ nm, a Gaussian pulse of 2 fs duration ($\Delta\tau$) at full-width-at half-maximum has a minimal bandwidth ($\Delta\lambda$) of 52.08 nm. Such an extension of the UV continuum is suitable for a number of time-resolved methods, including notably multi-dimensional spectroscopy.

An important case is the one of two-dimensional (2D) spectroscopy, allowing the extraction of complete information from a third-order nonlinear process. Briefly, the pump-probe scheme involves three pulses with distinct time delays (figure 5.9): a first pump pulse (k_1) at a frequency ω_1 creates a coherence between given states, a second pulse (k_2) is employed to induce a population transfer at a time t_1. The system is left to evolve and a probe pulse (k_3) at a time t_2 triggers the emission of a third-order nonlinear signal (k_{sig}) at a frequency ω_3. The signal is recorded by a spectrometer as a function of t_2. In order to extract both the amplitude and phase of the nonlinear response, k_{sig} can also be heterodyned with a reference signal, called 'local oscillator' (k_{LO}). By Fourier transform with respect to t_1, a 2D map depicting a series of spectrum for each pump frequency ω_1 can be obtained and a correlation between the pump and probe frequency is thus retrieved. An example of a 2D map is shown in figure 5.10(a).

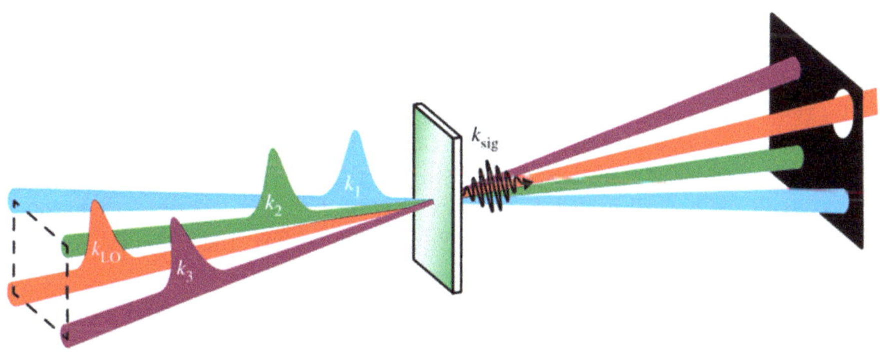

Figure 5.9. Pulses sequence in 2D spectroscopy. The 'boxcar' geometry is shown. Reproduced from [60] with permission of The Royal Society of Chemistry.

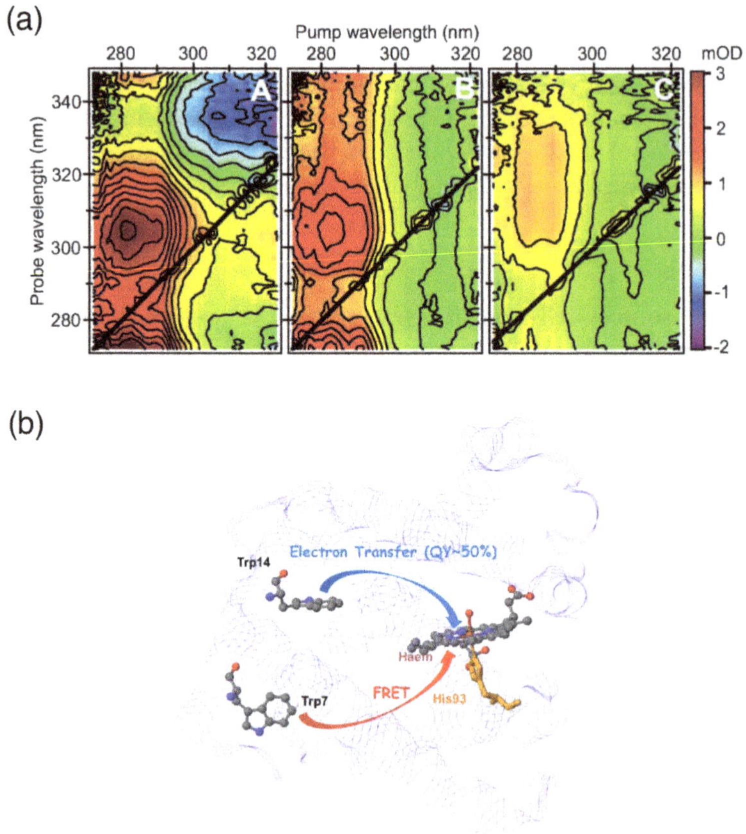

Figure 5.10. (a) 2D UV spectra of myoglobin for pump-probe delay $\tau_2 = 2.5$ ps (A), $\tau_2 = 17$ ps (B), $\tau_2 = 200$ ps (C). The black line shows the diagonal of the correlation spectra. The time resolution of the experiment is 150 fs. Reprinted from [62] with permission from AAAS. (b) Scheme of the molecular processes at play: While Trp7 undergoes the FRET mechanism (see main text), Trp14 undergoes a partial electron transfer process. Adapted from [66], with the permission of AIP Publishing. © 2019.

The additional dimensionality of the technique, compared, e.g., to 1D transient absorption, is that it not only enables tracking of the population and depletion of the involved states, but also describes the dynamical couplings between these states. Initially, the method has been used with low energy infrared pulses (2DIR) to study correlated vibrational dynamics, allowing one, for instance, to investigate unfolding dynamics in proteins [61]. With access to shorter (more energetic) wavelengths in the visible as well as broadband UV spectra, it has become possible to extend the method to 2D electronic spectroscopy (2DES), including the 2D UV scheme for which only a few experiments have been reported so far.

A seminal example is the one by Chergui and coworkers that identified an electron transfer mechanism in myoglobin [62]. The experiment consisted in tracking the UV-induced dynamical response of tryptophan (Trp) chromophores within the protein. Indeed, tryptophan is known to be an efficient probe to monitor the structure of biological functions via a dipole–dipole energy transfer process known as fluorescence-resonance-energy transfer (FRET) [63]. However, figure 5.10(a) shows the time-resolved 2D UV transient spectra obtained from the experiment, which identified a new electron transfer process between the Trp14 chromophore and the heme moiety of myoglobin, competing with the FRET process. Instead Trp7 was found to solely undergo FRET (figure 5.10(b)). The work provided one of the first real-time tracking experiments of long-range electron transfer in proteins, highlighting the capabilities of 2D UV spectroscopy. Other works exploiting the bandwidth of even broader UV pulses (as short as 20 fs) also include the investigation of the ultrafast relaxation response of the nucleobases adenine and uracil [64, 65].

Finally, electronic-vibrational spectroscopy (2DEV) is an extension of 2D electronic spectroscopy, in which a pump pulse triggers an electronic transition and a (lower frequency) probe pulse interrogates vibrational states (for a review on the various 2D spectroscopic techniques see [60]). By combining few-femtosecond UV-pump pulses with infrared probe pulses, this multi-dimensional technique promises to resolve the interaction between nuclear and electronic degrees of freedoms near conical intersections with an unprecedented amount of information.

5.3.3 Polarization state and future prospects

An important property of several biomolecular systems is their chiral character, meaning that they cannot be superimposed to their mirror image. Chiral molecules play a major role in pharmacology and biochemistry because each enantiomer (i.e., mirror image) can have completely different reactivities towards a chiral bioreceptor. The chiral character of molecules also provides them with optical activity, which can be exploited for circular dichroism measurements, where the absorption difference between left-handed (LCP) and right-handed circularly polarized (RCP) light of a chiral sample is evaluated (ΔA = LCP-RCP). In the UV spectral range near 200 nm, this powerful (static) technique is routinely employed in order to characterize the structure of proteins and identify conformational changes, leading to denaturation for instance [67]. The development of time-resolved analogs of circular dichroism in the UV is in its infancy. One of the main technical challenges responsible for this is that the design of ultrabroadband dispersion-free optics to control the polarization state of femtosecond UV pulses is extremely challenging.

Oppermann *et al* recently reported an instrument for time-resolved circular dichroism (TR-CD) experiments on liquid samples with a 500-fs temporal resolution [68]. For such measurements, circular dichroism spectra in the UV are acquired as a function of the time delay with respect to linearly polarized femtosecond pump pulses centered either at 400 or 266 nm. The probe spectrum ranges between 250 and 370 nm, which is suitable not only for the study of a number of dynamics in biological systems but also for several processes in organometallic complexes. An extension of the method to the characteristic probe region below 250 nm where the secondary structures of proteins absorb, such as α-helices and β-sheets [67], has not been reached yet but will allow the study of their structural changes in real time.

Alternatively, the quasi dispersion-free gas phase provides access to the highest achievable temporal resolution to track ultrafast dynamics in chiral molecules, both at the few-femtosecond and attosecond timescales. However, the density of the gaseous target is low compared to liquid samples of high concentration. This prevents the study of gaseous chiral molecules using time-resolved circular dichroism, whereas photoelectron circular dichroism (PECD) is an emerging technique that promises to tackle this challenge. In PECD, photoionization of a chiral molecule by circularly polarized light results in the emission of an electron either in the backward or forward direction with respect to the light propagation axis. The emission direction depends on both the light helicity and the handedness of the enantiomer, and exchanging one of them reverses the photoelectron emission direction. Typically, the differential photoelectron angular distribution between LCP and RCP light serves as an observable (see figure 5.11(b), bottom row), and is acquired using a velocity map imaging (VMI) spectrometer [69]. The static version of this technique has shown extreme sensitivity to molecular properties, being able to resolve orbitals and vibrational levels [70, 71], as well as conformers and isomers [72, 73]. The first time-resolved PECD (TR-PECD) experiment was achieved in 2016 by Comby *et al*, shown in figure 5.11(a). Linearly polarized UV-pump pulses (201 nm) were employed to excite fenchone molecules and the photoelectron angular distribution was evaluated with both LCP and RCP ionizing probe pulses [74].

Figure 5.11(b) (bottom row) shows a series of 'difference' images (LCP-RCP) containing the signature of an electronic relaxation process that was identified to

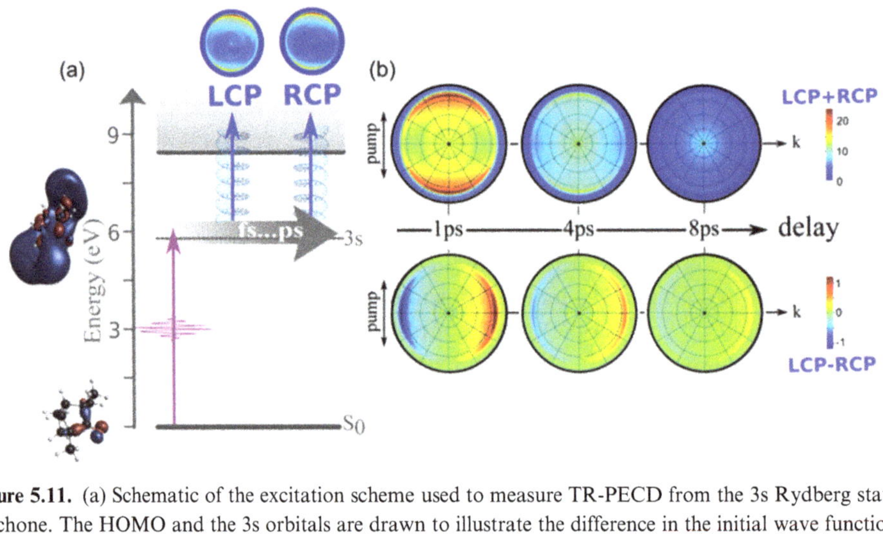

Figure 5.11. (a) Schematic of the excitation scheme used to measure TR-PECD from the 3s Rydberg state of fenchone. The HOMO and the 3s orbitals are drawn to illustrate the difference in the initial wave function of the outgoing electron. (b) The photoelectron images recorded on (1R)-(-)-fenchone ($e = -1$) for each probe helicity (LCP-p = 1 or RCP-p = -1) are subtracted one from the other (LCP – RCP), while the sum of these images (LCP + RCP) provides the photoelectron spectrum. These (LCP – RCP) and (LCP + RCP) images, called 'difference' and 'sum' images, respectively, are shown for three pump-probe delays. Note that the two photon beams are quasi-copropagating with the polarization axis parallel to the detector. Adapted from [74]. © 2016 ACS Publications.

take place with a time constant of about 3.3 ps. In order to enhance the temporal resolution of such a method to the few-femtosecond regime, control over the polarization state of the recently reported ultrabroadband UV pulses is necessary [15, 36]. This will notably enable the study of ultrafast dynamics in bio-relevant chiral molecules such as amino-acids for instance. As previously mentioned, starting from linearly polarized few-cycle UV pulses, the design of optics with minimal dispersion is extremely challenging. However, circularly polarized femtosecond pulses in the deep ultraviolet were recently obtained via seeded degenerate four-wave mixing in stretched hollow-core fibers filled with gas [39]. With a pulse energy at the microjoule level, the UV spectra supports a transform-limit duration of 4 fs. Such a technological advance will surely open the way to several experiments requiring circularly polarized UV radiation with extreme temporal resolution.

5.4 Conclusion

The development of ultrafast UV light sources providing femtosecond pulses in the few-cycle regime is in demand for engineering new technologies as well as for fundamental research. This is easily justified since ultraviolet radiation is involved in a wide range of scientific areas, including energy-harvesting, atmospheric chemistry, and biochemistry. While mature time-resolved spectroscopic techniques in molecular science have allowed the tracking in real time of a number of photochemical and photobiological ultrafast phenomena, they have barely been applied to the UV spectral domain due to the technical challenges for the generation and manipulation of ultrabroadband UV pulses.

The state-of-the-art methods for the generation of few-femtosecond UV pulses have been highlighted in this chapter, and their advantages and limitations have been detailed. For parametric processes in crystals, the limited thickness allowed in order to prevent dispersion imposes to manipulate at best the bandwidth and spectral phase of the generated UV pulses to obtain an optimal compression. Advanced techniques have allowed sub-10-fs (with a lower limit of 7.1 fs) pulses to be obtained with energies up to the microjoule level. Instead, gas media allows one to reach much shorter pulse durations. THG of near-infrared pulses at 800 nm can be obtained in gas cells filled with noble gas. In this case, sub-2-fs UV pulses have been demonstrated experimentally, and the production of pulses in the attosecond time domain is expected to be reached. Such an approach, however, requires the cell to handle several bars of gas pressure in order to obtain UV pulse energies at the nanojoule level. An alternative method using gas media consists in exploiting nonlinear processes during light propagation in photonic crystal fibers or capillaries where phase-matching conditions can be further optimized. Such emerging generation schemes allowed 3 fs UV pulses up to the microjoule energy level. Moreover control over the polarization state, that is a currently a severe technological limitation, has just been demonstrated using capillaries.

The unprecedented temporal resolution and bandwidth that provide few-femtosecond UV pulses for UV-based pump-probe measurements opens ways for a wide range of experiments in molecular science. This will notably allow the investigation of rapid structural changes in molecules as well as electron motions on timescales from few tens of femtoseconds down to the attosecond, with a particular interest in systems that are essential to life including ozone, amino-acids, nucleobases, and the corresponding macromolecules such as proteins and DNA strands. Future development of few-cycle UV pulses using high power, high repetition rate laser systems (>100 kHz) based on rare-earth doped gain media will also enable the success of experiments currently affected by poor statistics despite long acquisition times. There is no doubt that several advancements will be realized due to this 'exciting' spectral range.

References

[1] Laurens H 1928 The physiological effects of radiation *Physiol. Rev.* **8** 1–91

[2] Crowther J 2019 UV reflectance photography of skin–what are you imaging? *Int. J. Cosmet. Sci.* **42** 11

[3] Hearn R M R, Kerr A C, Rahim K F, Ferguson J and Dawe R S 2008 Incidence of skin cancers in 3867 patients treated with narrow-band ultraviolet B phototherapy *Br. J. Dermatol.* **159** 931–5

[4] Wang X, Jiang M, Zhou Z, Gou J and Hui D 2017 3D printing of polymer matrix composites: a review and prospective *Composites* B **110** 442–58

[5] Wu B and Kumar A 2007 Extreme ultraviolet lithography: a review *J. Vac Sci. Technol.* B **25** 1743–61

[6] Savage N 2007 Ultraviolet lasers *Nat. Photon.* **1** 83–5

[7] Franken P A, Hill A E, Peters C W and Weinreich G 1961 Generation of optical harmonics *Phys. Rev. Lett.* **7** 118–9

[8] Hockberger P E 2002 A history of ultraviolet photobiology for humans, animals and microorganisms *Photochem. Photobiol.* **76** 561–79

[9] Engelsen O 2010 The relationship between ultraviolet radiation exposure and vitamin D status *Nutrients* **2** 482–95

[10] Sinha R P and Häder D-P 2002 UV-induced DNA damage and repair: a review *Photochem. Photobiol. Sci.* **1** 225–36

[11] Voet D, Gratzer W B, Cox R A and Doty P 1963 Absorption spectra of nucleotides, polynucleotides, and nucleic acids in the far ultraviolet *Biopolymers* **1** 193–208

[12] Ravanat J-L, Douki T and Cadet J 2001 Direct and indirect effects of UV radiation on DNA and its components *J. Photochem. Photobiol.* B **63** 88–102

[13] Gross A J and Herrmann T R W 2007 History of lasers *World J. Urol.* **25** 217–20

[14] Basting D, Pippert K and Stamm U 2002 History and future prospects of excimer lasers *Second Int. Symp. on Laser Precision Microfabrication* vol 4426

[15] Galli M *et al* 2019 Generation of deep ultraviolet sub-2-fs pulses *Opt. Lett.* **44** 1308–11

[16] McPherson A, Gibson G, Jara H, Johann U, Luk T S, McIntyre I A, Boyer K and Rhodes C K 1987 Studies of multiphoton production of vacuum-ultraviolet radiation in the rare gases *J. Opt. Soc. Am.* B **4** 595–601

[17] Motz H 1951 Applications of the radiation from fast electron beams *J. Appl. Phys.* **22** 527–35

[18] Brabec T and Krausz F 1997 Nonlinear optical pulse propagation in the single-cycle regime *Phys. Rev. Lett.* **78** 3282–5

[19] Seka W, Jacobs S D, Rizzo J E, Boni R and Craxton R S 1980 Demonstration of high efficiency third harmonic conversion of high power Nd-glass laser radiation *Opt. Commun.* **34** 469–73

[20] Beutler M, Ghotbi M, Noack F, Brida D, Manzoni C and Cerullo G 2009 Generation of high-energy sub-20 fs pulses tunable in the 250-310 nm region by frequency doubling of a high-power noncollinear optical parametric amplifier *Opt. Lett.* **34** 710

[21] Baum P, Lochbrunner S and Riedle E 2004 Generation of tunable 7-fs ultraviolet pulses: achromatic phase matching and chirp management *Appl. Phys.* B **79** 1027–32

[22] Baum P, Lochbrunner S and Riedle E 2004 Tunable sub-10-fs ultraviolet pulses generated by achromatic frequency doubling *Opt. Lett.* **29** 1686

[23] Varillas R B, Candeo A, Viola D, Garavelli M, De Silvestri S, Cerullo G and Manzoni C 2014 Microjoule-level, tunable sub-10 fs UV pulses by broadband sum-frequency generation *Opt. Lett.* **39** 3849–52

[24] Tan H-S, Schreiber E and Warren W S 2002 High-resolution indirect pulse shaping by parametric transfer *Opt. Lett.* **27** 439

[25] Polyanskiy M N 2021 Refractive index database https://refractiveindex.info (Accessed 2021-02-02)

[26] Lehmeier H J, Leupacher W and Penzkofer A 1985 Nonresonant third order hyperpolarizability of rare gases and N_2 determined by third harmonic generation *Opt. Commun.* **56** 67–72

[27] Banks P S, Feit M D and Perry M D 1998 High intensity direct third harmonic generation in BBO *Nonlinear Optics'98. Materials, Fundamentals and Applications Topical Meeting (Cat. No. 98CH36244)* (IEEE: Piscataway, NJ) pp 268–70

[28] Gubler U and Bosshard C 2000 Optical third-harmonic generation of fused silica in gas atmosphere: absolute value of the third-order nonlinear optical susceptibility χ (3) *Phys. Rev.* B **61** 10702

[29] Graf U, Fieß M, Schultze M, Kienberger R, Krausz F and Goulielmakis E 2008 Intense few-cycle light pulses in the deep ultraviolet *Opt. Express* **16** 18956–63

[30] Reiter F, Graf U, Serebryannikov E E, Schweinberger W, Fieß M, Schultze M, Azzeer A M, Kienberger R, Krausz F and Zheltikov A M *et al* 2010 Route to attosecond nonlinear spectroscopy *Phys. Rev. Lett.* **105** 243902

[31] Vishnubhatla K C, Bellini N, Ramponi R, Cerullo G and Osellame R 2009 Shape control of microchannels fabricated in fused silica by femtosecond laser irradiation and chemical etching *Opt. Express* **17** 8685–95

[32] Ciriolo A G, Vázquez R M, Roversi A, Frezzotti A, Vozzi C, Osellame R and Stagira S 2020 Femtosecond laser-micromachining of glass micro-chip for high order harmonic generation in gases *Micromachines* **11** 165

[33] Durfee C G III, Rundquist A R, Backus S, Herne C, Murnane M M and Kapteyn H C 1999 Phase matching of high-order harmonics in hollow waveguides *Phys. Rev. Lett.* **83** 2187

[34] Durfee C G, Backus S, Kapteyn H C and Murnane M M 1999 Intense 8-fs pulse generation in the deep ultraviolet *Opt. Lett.* **24** 697

[35] Misoguti L, Backus S, Durfee C G, Bartels R, Murnane M M and Kapteyn H C 2001 Generation of broadband VUV light using third-order cascaded processes *Phys. Rev. Lett.* **87** 013601

[36] Brahms C, Austin D R, Tani F, Johnson A S, Garratt D, Travers J C, Tisch J W G, Russell P S J and Marangos J P 2019 Direct characterization of tuneable few-femtosecond dispersive-wave pulses in the deep UV *Opt. Lett.* **44** 731–4

[37] Travers J C, Grigorova T F, Brahms C and Belli F 2019 High-energy pulse self-compression and ultraviolet generation through soliton dynamics in hollow capillary fibres *Nat. Photon.* **13** 547–54

[38] Brahms C, Belli F and Travers J C 2020 Resonant dispersive wave emission in hollow capillary fibers filled with pressure gradients *Opt. Lett.* **45** 4456
[39] Lekosiotis A, Belli F, Brahms C and Travers J C 2020 Generation of broadband circularly polarized deep-ultraviolet pulses in hollow capillary fibers *Opt. Lett.* **45** 5648
[40] Belli F, Abdolvand A, Chang W, Travers J C and Russell P St 2015 Vacuum-ultraviolet to infrared supercontinuum in hydrogen-filled photonic crystal fiber *Optica* **2** 292
[41] Belli F, Abdolvand A, Travers J C and Russell P St 2019 Highly efficient deep UV generation by four-wave mixing in gas-filled hollow-core photonic crystal fiber *Opt. Lett.* **44** 5509
[42] Trebino R 2000 *Frequency-Resolved Optical Gating: The Measurement of Ultrashort Laser Pulses: The Measurement of Ultrashort Laser Pulses* (Berlin: Springer)
[43] Iaconis C and Walmsley I A 1998 Spectral phase interferometry for direct electric-field reconstruction of ultrashort optical pulses *Opt. Lett.* **23** 792–4
[44] Birge J R, Ell R and Kärtner F X 2006 Two-dimensional spectral shearing interferometry for few-cycle pulse characterization *Opt. Lett.* **31** 2063–5
[45] Li M, Nibarger J P, Guo C and Gibson G N 1999 Dispersion-free transient-grating frequency-resolved optical gating *Appl. Opt.* **38** 5250–3
[46] Witting T, Greening D, Walke D, Matia-Hernando P, Barillot T, Marangos J P and Tisch J W G 2016 Time-domain ptychography of over-octave-spanning laser pulses in the single-cycle regime *Opt. Lett.* **41** 4218–21
[47] Reiter F, Graf U, Schultze M, Schweinberger W, Schröder H, Karpowicz N, Azzeer A M, Kienberger R, Krausz F and Goulielmakis E 2010 Generation of sub-3 fs pulses in the deep ultraviolet *Opt. Lett.* **35** 2248–50
[48] Kida Y 2020 Transient grating in a thin gas target for characterization of extremely short optical pulses *Opt. Lett.* **45** 2231–4
[49] Hazra C, Samanta T and Mahalingam V 2014 A resonance energy transfer approach for the selective detection of aromatic amino acids *J. Mater. Chem.* C **2** 10157–63
[50] Joshi D, Kumar D, Maini A K and Sharma R C 2013 Detection of biological warfare agents using ultra violet-laser induced fluorescence LIDAR *Spectrochim. Acta* A **112** 446–56
[51] Calegari F, Sansone G, Stagira S, Vozzi C and Nisoli M 2016 Advances in attosecond science *J. Phys. B: At. Mol. Opt. Phys.* **49** 062001
[52] Attar A R, Bhattacherjee A, Pemmaraju C D, Schnorr K, Closser K D, Prendergast D and Leone S R 2017 Femtosecond x-ray spectroscopy of an electrocyclic ring-opening reaction *Science* **356** 54–9
[53] Latka T *et al* 2019 Femtosecond wave-packet revivals in ozone *Phys. Rev.* A **99** 063405
[54] Chang K F, Reduzzi M, Wang H, Poullain S M, Kobayashi Y, Barreau L, Prendergast D, Neumark D M and Leone S R 2020 Revealing electronic state-switching at conical intersections in alkyl iodides by ultrafast XUV transient absorption spectroscopy *Nat. Commun.* **11** 4042
[55] Woodward R B and Hoffmann R 1965 Stereochemistry of electrocyclic reactions *J. Am. Chem. Soc.* **87** 395–7
[56] Hoffmann R 1968 Trimethylene and the addition of methylene to ethylene *J. Am. Chem. Soc.* **90** 1475–85
[57] Remacle F and Levine R D 2006 An electronic time scale in chemistry *Proc. Natl Acad. Sci. USA* **103** 6793–8
[58] Calegari F *et al* 2014 Ultrafast electron dynamics in phenylalanine initiated by attosecond pulses *Science* **346** 336–9

[59] Kraus P M *et al* 2015 Measurement and laser control of attosecond charge migration in ionized iodoacetylene *Science* **350** 790–5
[60] Oliver T A A 2018 Recent advances in multidimensional ultrafast spectroscopy *R. Soc. Open Sci.* **5** 171425
[61] Hoi S C, Ganim Z, Jones K C and Tokmakoff A 2007 Transient 2D IR spectroscopy of ubiquitin unfolding dynamics *Proc. Natl Acad. Sci. USA* **104** 14237–42
[62] Consani C, Auböck G, van Mourik F and Chergui M 2013 Ultrafast tryptophan-to-heme electron transfer in myoglobins revealed by UV 2D spectroscopy *Science* **339** 1586–90
[63] Lipman E A, Schuler B, Bakajin O and Eaton W A 2003 Single-molecule measurement of protein folding kinetics *Science* **301** 1233–5
[64] West B A, Womick J M and Moran A M 2011 Probing ultrafast dynamics in adenine with mid-UV four-wave mixing spectroscopies *J. Phys. Chem.* A **115** 8630–7
[65] Tseng C H, Sándor P, Kotur M, Weinacht T C and Matsika S 2012 Two-dimensional Fourier transform spectroscopy of adenine and uracil using shaped ultrafast laser pulses in the deep UV *J. Phys. Chem.* A **116** 2654–61
[66] Chergui M 2019 Ultrafast molecular photophysics in the deep-ultraviolet *J. Chem. Phys.* **150** 070901
[67] Greenfield N J 2007 Using circular dichroism spectra to estimate protein secondary structure *Nat. Protocols* **1** 2876–90
[68] Oppermann M, Bauer B, Rossi T, Zinna F, Helbing J, Lacour J and Chergui M 2019 Ultrafast broadband circular dichroism in the deep ultraviolet *Optica* **6** 56–60
[69] Eppink A T J and Parker D H 1997 Velocity map imaging of ions and electrons using electrostatic lenses: application in photoelectron and photofragment ion imaging of molecular oxygen *Rev. Sci. Instrum.* **68** 3477–84
[70] Nahon L, Garcia G A, Harding C J, Mikajlo E and Powis I 2006 Determination of chiral asymmetries in the valence photoionization of camphor enantiomers by photoelectron imaging using tunable circularly polarized light *J. Chem. Phys.* **125** 114309
[71] Garcia G A, Nahon L, Daly S and Powis I 2013 Vibrationally induced inversion of photoelectron forward-backward asymmetry in chiral molecule photoionization by circularly polarized light *Nat. Commun.* **4** 2132
[72] Turchini S, Catone D, Contini G, Zema N, Irrera S, Stener M, Di Tommaso D, Decleva P and Prosperi T 2009 Conformational effects in photoelectron circular dichroism of alaninol *Chem. Phys. Chem.* **10** 1839–46
[73] Powis I, Harding C J, Garcia G A and Nahon L 2008 A valence photoelectron imaging investigation of chiral asymmetry in the photoionization of fenchone and camphor *Chem. Phys. Chem.* **9** 475–83
[74] Comby A *et al* 2016 Relaxation dynamics in photoexcited chiral molecules studied by time-resolved photoelectron circular dichroism: toward chiral femtochemistry *J. Phys. Chem. Lett.* **7** 4514–9

Lightning Source UK Ltd.
Milton Keynes UK
UKHW051615250921
391114UK00003B/69

9 780750 325349